CW00543837

HANDBOOK OF MEAT PRODUCT TECHNOLOGY

HANDBOOK OF MEAT PRODUCT TECHNOLOGY

M.D. RANKEN

BSc Tech, CChem, MChemA, MFC,
FRSC, FIFST, FInstM
Consultant Food Technologist

b

Blackwell
Science

Editorial Offices:
Osney Mead, Oxford OX2 0EL
25 John Street, London WC1N 2BL
23 Ainslie Place, Edinburgh EH3 6AJ
350 Main Street, Malden
MA 02148 5018, USA
54 University Street, Carlton
Victoria 3053, Australia
10, rue Casimir Delavigne
75006 Paris, France

Other Editorial Offices:

Blackwell Wissenschafts-Verlag GmbH
Kurfürstendamm 57
10707 Berlin, Germany

Blackwell Science KK
MG Kodenmacho Building
7–10 Kodenmacho Nihombashi
Chuo-ku, Tokyo 104, Japan

First published 2000

Set in 10/12pt Times
by DP Photosetting, Aylesbury, Bucks
Printed and bound in Great Britain by
MPG Books Ltd, Bodmin, Cornwall

DISTRIBUTORS

Marston Book Services Ltd
PO Box 269
Abingdon
Oxon OX14 4YN
(*Orders:* Tel: 01235 465500
Fax: 01235 465555)

USA
Blackwell Science, Inc.
Commerce Place
350 Main Street
Malden, MA 02148 5018
(*Orders:* Tel: 800 759 6102
781 388 8250
Fax: 781 388 8255

Canada
Login Brothers Book Company
324 Saulteaux Crescent
Winnipeg, Manitoba R3J 3T2
(*Orders:* Tel: 204 837-2987
Fax: 204 837-3116)

Australia
Blackwell Science Pty Ltd
54 University Street
Carlton, Victoria 3053
(*Orders:* Tel: 03 9347 0300
Fax: 03 9347 5001)

A catalogue record for this title is available from the British Library

ISBN 0-632-05377-1

Library of Congress
Cataloging-in-Publication Data
Ranken, Michael D.
Handbook of meat product technology/
M.D. Ranken.
p. cm.
Includes bibliographical references and index.
ISBN 0-632-05377-1 (pbk.)
1. Slaughtering and slaughter-houses.
2. Meat. I. Title.

TS 1960. R25 2000
664.9—dc21 99-059649

For further information on
Blackwell Science, visit our website:
www.blackwell-science.com

Contents

Preface

The need has been evident for some time for a short handbook of meat product technology, expressed in the language and style of modern science and technology. The majority of the older texts are written by and in the style of the private or small-scale butcher with good practical experience, knife in hand, of meat animals and meat. They give practical instruction on the employment of the butcher's practical understanding to make the best commercial use of the meat which he has before him. But the present-day college or university graduate, appointed to manage a factory department or a complete meat factory, where all of those skills may be employed on the large scale and subdivided among different groups of people, does not usually himself (or, often enough, herself) enjoy the same practical skills and experience. His (or her) skills are science-based and hence there is a need to understand the meat technology from that scientific basis.

This handbook is an attempt to meet that need. It attempts to give proper weight on the one hand to the craft element in all modern and traditional meat product manufacturing and on the other hand to the scientific reasons (so far as they are known) for those practices, their best results and their occasional failures.

NOTE

Reference is made occasionally in the text to particular machines, ingredients, etc., giving the names of manufacturers or brands of which the writer has knowledge or experience. In no case should any such reference by name to a material, ingredient, machine or process be taken to imply endorsement of the named product over any similar product.

M.D. Ranken
Hythe, January 2000

Acknowledgements

I must express deep gratitude to the Leatherhead Food Research Association for all the opportunities which I had when I was there from 1970 until 1984, to formulate the ideas and to discover the information on which this handbook is based. I include warm thanks to my colleagues in the Meat and Fish Products Laboratory at that time for their contributions to the work and the thinking which we did together. Those former colleagues include, particularly, Percy Barnet, Dr Gar Evans, Mats Henriques, Anne Manning (then Anne Lewin), Gilbert Oliphant, Dr Sue Valentine (then Sue Richards) and the late Dr John Wood.

I have been much influenced by two books which are standard works in the field: Frank Gerrard's *Sausage and Small Goods Production* and the successive editions of Ralston Lawrie's *Meat Science*. I am very grateful for all that they have taught me.

Figures 9.1 and 9.2 are from *Food Industries Manual*, 25th edition, with permission of the Editors. Figure 1.4 is reproduced from *Food Standards Committee Report on Meat Products 1980*, with permission of the Controller of The Stationery Office, London. Figure 8.1 is reproduced by permission of Urschel International Ltd, Leicester. Figure 8.4 is by permission of DMV bv, Veghel, The Netherlands.

I owe thanks also to a number of people now more closely associated with the meat products industry than I, who have given generously of their time and knowledge in checking that the information here is accurate and up to date: Professor Keith Anderson, University of North London; Professor Joe Buckley, University College, Cork; Dr Ron Kill, Micron Laboratories; Professor David Ledward, Reading University; Mr Fred Mallion, Worshipful Company of Butchers; Professor Geoff Mead, Royal Veterinary College; Mr Michael Nightingale and Mr Nick Church, Fibrisol Ltd; Dr Robert Shaw, Campden and Chorleywood Food Research Association; Dr Tom Toomey, Ventress Technical Services Ltd; and Dr Jean-Luc Vendeuvre, Centre Technique de la Salaison, Charcuterie et Conservation de la Viande, Maisons Alfort, France.

Finally, to all the other people whom I no longer properly remember who have contributed materials and ideas, often no doubt unwittingly, to this handbook, my apologies but many thanks.

Part One
PRINCIPLES

1 Manufacturing Meat

SOURCES

The main sources of manufacturing meats are noted below.

Cattle

Cow

'Cow beef' comes from animals at the end of their useful period of milk production, usually 5–8 years old at slaughter, when milk yield has begun to fall (but note that Regulations in the UK relating to BSE prohibit the use of meat from cattle more than 30 months old).

- Dairy breeds in the UK include Ayrshire, Jersey, Friesian and Holstein; these are not usually very meaty.
- 'Dual purpose' breeds such as Dexter, Red Poll and Shorthorn have better conformation and meat yield.

Calf

Milk production is a sequel to calving; male calves and many of the females are surplus to the requirement for new milk cattle. The surplus provides **veal** calves grown to 3–4 months (bobby calf, bobby veal – up to 3 months old).

Steer, bullock

Males grown to meat weights (450 kg live weight or more); not usually available for manufacturing, except that in large meat plants they may provide trimmings and the less 'noble' cuts (see below).

- 'Dual purpose' breeds, as above, are intended to give females with good milk production and males with good meat yield.
- 'Heavy' breeds (e.g. Aberdeen Angus, Hereford, Charolais), in which both males and females are grown for meat.
- Beef × dairy crosses constitute the bulk of beef slaughtered in the UK.

Pig

Bacon pig

Bred for long backs with low to moderate fat cover. Convert economically to bacon.

Heavy hog

Fatter and proportionately shorter. Bred for high growth rate and good feed conversion. Used for mixed manufacture – fresh pork, some bacon, some sausage and pie meat.

'Continental' breeds

Tendency to even more fat than the heavy hog.

Boars

The meat of boars which have been used for breeding is usually strongly tainted with the smell of male sex hormone ('boar taint'). Young, uncastrated males, slaughtered at or before puberty, have good feed conversion and conformation and may be free from this taint.

Sheep

Differences of breed or sex are not significant for manufacturing purposes, except that 'ram taint' may be encountered in the meat of old breeding rams.

Poultry

Chicken

- Hens from egg production ('spent' hens). Usually about 18 months old, small, and with relatively poor conformation and meat yield. They are cheap and are the main source of manufacturing chicken meat.
- Broilers. The term means suitable for grilling; in the USA it covers birds up to about 1.5 kg dressed weight, but in the UK heavier birds are included, up to 3–4 kg. Age 6–10 weeks.
- Broiler breeder hens. These are the parents of broilers after their productive egg-laying life. They are larger and have better meat yield than normal egg-layers. Numbers available are relatively small.

Note that:

- Flavour is stronger but texture is tougher in the older birds (hens versus broilers).
- A small proportion of birds may be grown under 'free range' conditions, some laying birds in 'pole barns', etc., but almost all the remainder are grown intensively, broilers in large open houses, hens in cages.
- Flavour and texture differences due to breed or growing conditions are negligible.

Turkey

For domestic sale, carcasses may range down to 2–3 kg dressed weight. For manufacturing, mainly male birds are used; these are usually 10–15 kg dressed weight.

Other species

From time to time there may be interest in ensuring that species such as horse, hippopotamus or kangaroo are not supplied fraudulently in place of beef. These and other species may be of interest for pet foods, depending on availability and price.

MANUFACTURING CUTS

'Noble' and 'less noble' meats

The 'noble' cuts are those most highly regarded by chefs and gourmets because they have:

- high contents of muscle
- small amounts of fat, which is on the outside of the meat and so can be easily removed if unwanted
- low contents of connective tissue or gristle and none in the form of large, thick pieces
- small amounts of bone, which can be easily removed.

Meat with these properties is:

- simple to cook, e.g. by grilling or roasting
- tender when lightly cooked
- simple to serve and provides large portions consisting almost entirely of desirable lean meat
- highly regarded and therefore highly priced.

'Noble' cuts come from the hindquarter of the animal where there are:

- fewer moving parts
- simpler bone structure
- a few large muscles
- less connective tissue
- fat deposits mainly on the outside.

The less 'noble' cuts have the reverse characteristics from those listed above and are more likely to be used for manufacturing. They come mainly from the forequarter, where there are:

- many and complex moving parts
- a complex bone structure
- many, smaller muscles
- more connective tissue.

Cuts from the belly or flank, where there are no bones to give support in the live animal, have particularly strong (therefore tough) connective tissues; there are also more or less thick layers of fat between the muscles.

Figure 1.1 shows how some of these factors apply in typical cuts of beef.

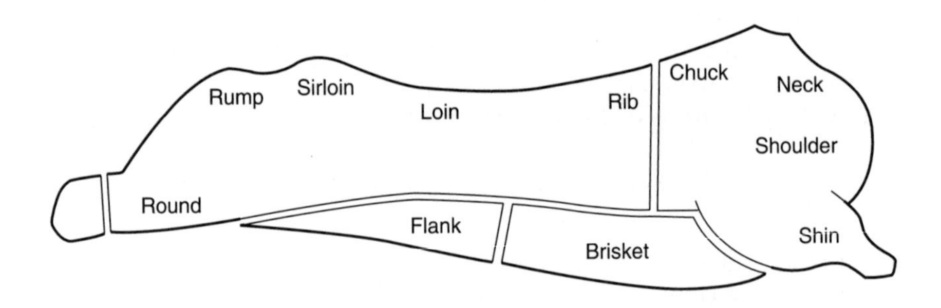

	Approx. %			
	Muscle	Fat	Connective tissue	Bone
Round	59	5	11	15
Brisket	48	18	17	17
Neck	48	8	19	25
Flank	58	17	25	0

Fig. 1.1 Composition of typical cuts of beef.

'Primal cuts' of beef as supplied for manufacturing commonly consists of the forequarter (chuck + neck + shoulder + shin) or 'Pistola fores' (forequarter + brisket + some flank).

Figure 1.2 shows the major cuts of pork.

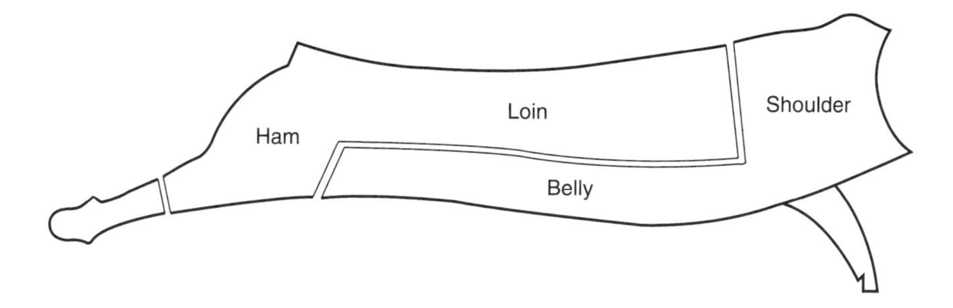

Fig. 1.2 Main cuts of pork

Reasons for making meat products

A major purpose for converting meat into meat products must be **to modify or upgrade the less noble cuts of meat**, together with any edible trimmings of fat and connective tissue removed from the more noble cuts; and to make the flavour and texture more acceptable to consumers than they would be if treated only by the simple cooking and serving methods which are appropriate for the noble cuts.

The technical problems which must be dealt with in improving the acceptability of such meat are:

- to remove bones
- to make the connective tissues less objectionable
- to present the available fat in more acceptable form
- to leave flavour and nutritive value unimpaired or even improved.

An alternative purpose may be **to preserve the meat**. Here the question of the 'nobility' of the meat may not be so important. For bacon and ham manufacture, for instance, the pigs may be specially selected for properties relevant to the quality of the final product, such as back fat thickness or low tendency to PSE (see page 10); meat without these desirable properties may be diverted to other products or even to the butchery trade. Preservation may of course be undertaken in addition to the upgrading described above, e.g. the canning of luncheon meat. Preservation methods are dealt with in later chapters.

Economic factors may distort these purposes in special cases; for example, a sausage factory might be run at full capacity to satisfy an existing market, even at the cost of using more noble meat as raw material. Such distortions are likely to be temporary.

COMPONENTS OF MEAT AND THEIR PROPERTIES

Lean meat

Lean meat or muscle consists of:	%
• a **contractile mechanism** consisting of myofibrillar protein (actin, myosin, etc.), in the form of many fibrils, fibres and fibre bundles	10.0
• each encased in light tubing or **netting** (connective tissue), consisting of collagen and elastin	2.0
• surrounded by **fluid** (sarcoplasm), consisting of water (75.0%), sarcoplasmic protein (6.0%), and other soluble substances including myoglobin (red colour), salts, vitamins, etc.	84.5
• and some fat, sinews, **nerves, blood vessels**, etc.	3.5

The actin–myosin contractile system is represented simply in Fig. 1.3; a more detailed diagram of the structure of muscle is given in Fig. 1.4.

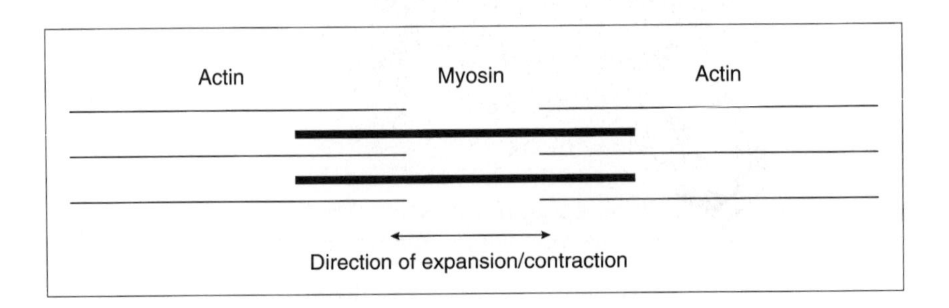

Fig. 1.3 The actin–myosin system.

Post-mortem changes

The chemical changes which take place on the death of the animal, changing muscle into meat, are complicated but fairly well understood. For detailed discussion see standard textbooks, e.g. R.A. Lawrie's *Meat Science*.

The most important changes are as follows.

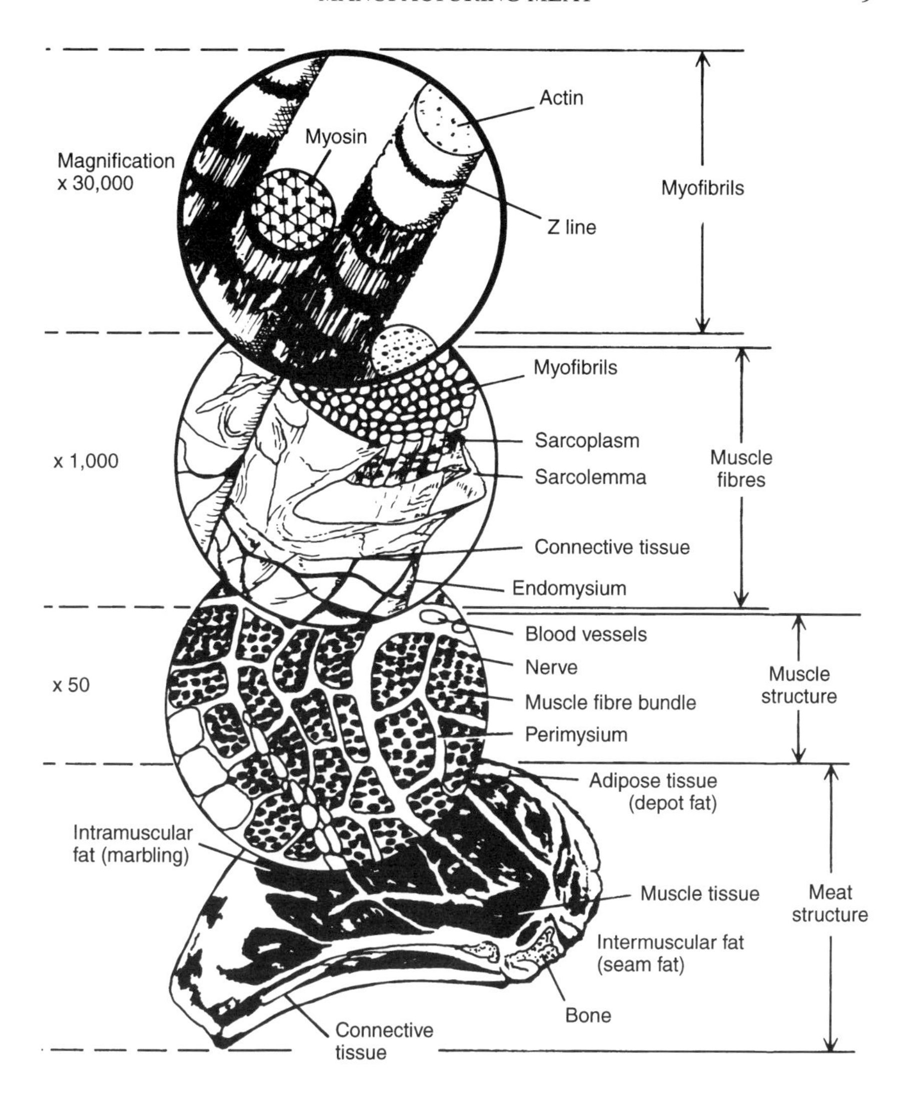

Fig. 1.4 Structure of meat at various magnifications.

(a) Effects related to pH

When normal metabolism and the supply of oxygen to the bloodstream cease, glycogen (the animal's energy supply, derived from food) is turned to lactic acid and the pH falls, normally from 7.0–7.2 to 5.5–6.5. This process is known as glycolysis. In abnormal cases the following conditions may occur.

- **PSE condition** – pale, soft, exudative (= wet) meat. If the pH falls very rapidly (glycogen supply being adequate) because of nervous excitement at the time of slaughter especially in stress-susceptible animals, (e.g. Pietrain or Danish Landrace pigs), the result is a low pH value (not abnormally low, but reached quickly while the carcass is still warm). This leads to precipitation of soluble proteins (sarcoplasmic protein), poor water binding and pale colour.
- **DFD condition** – dry, firm, dark meat. If the glycogen supply is low because of hunger (starvation), exercise (exhaustion), or long-term stress in the live animal, little lactic acid can be formed and the ultimate pH is high. This leads to deeper colour, closer texture and better water binding, but poorer microbiological quality. Other names for the condition are 'dark cutting' in beef and 'glazy' in bacon.

The avoidance of either of these conditions depends upon good conditions of transport, lairage and slaughter. The best quality meat comes therefore from healthy, well fed, unstressed animals.

(b) Effects related to rigor mortis

On the death of the animal the ATP (adenosine triphosphate) in the muscles goes to ADP (adenosine diphosphate) and AMP (adenosine monophosphate), with a release of energy which causes contraction, i.e. rigor mortis. After a time the muscles relax again, that is, there is resolution of rigor mortis.

Times for onset and resolution of rigor in different animals are as shown in Table 1.1.

Table 1.1 Times for onset and resolution of rigor

	Approximate time to onset of rigor	Approximate time for resolution of rigor
Cattle	12–24 h	2–6 d*
Pigs	6–12 h	1–3 d
Turkey	$\frac{1}{2}$–2 h	6–24 h
Chickens	$\frac{1}{2}$–1 h	4–6 h

* To 14 d for maximum tenderness.

In chilled meat supplied for manufacturing, these processes have normally been completed and no problems arise. But if the rigor-resolution sequence is interrupted by cutting, chilling, freezing or cooking, toughness may result.

- **Cutting**. Cutting before or during rigor allows muscles to shorten and may cause toughness when the meat is cooked.
- **Chilling or freezing**. Chilling the meat rapidly, immediately after slaughter (i.e. if the temperature falls to +10°C (50°F) by the time pH reaches *c*. 5.5 and rigor commences) leads to 'cold shortening' and tough meat. This is a problem with sheep and sometimes cattle, but not normally with pigs or poultry. The remedy is to chill more slowly or to use electrical stimulation: see below.

  Freezing the meat early in rigor or before rigor commences, when residual ATP is still present, leads to '**thaw rigor**': strong contraction with toughening when the meat is thawed. If the frozen meat is stored for a long time (months), the ATP gradually disappears and thaw rigor diminishes. If the meat is held at −5°C (23°F) for several hours before thawing, the chemical changes continue but the meat is unable to contract and therefore does not toughen.

  Freezing after rigor gives no special problems.
- **Cooking**. If this is done **before onset of rigor** (i.e. immediately after slaughter) it produces very tender meat (in theory: in practice it may not be possible to work fast enough and rigor will commence before or during cooking). Cooking **during rigor** results in tough meat. Cooking **after resolution of rigor** produces tender meat. Tenderness increases with time before cooking, up to a maximum.

The toughness of contracted muscles is probably due to a combination of two factors:

- compression of the muscle structure as the actomyosin system contracts, and
- contraction and tightening of the connective tissue network in the muscle sheaths.

Several methods are in use to reduce the toughening due to these effects:

- **The 'Tenderstretch' process**: the carcass is hung from the 'aitch' bone immediately after slaughter so that the maximum proportion of the 'noble' muscles are stretched, improving their tenderness.
- **'Hot' meat processing**. Salt treatment before rigor prevents contraction (although ATP is still lost), and gives meat with high water holding capacity.
- **Electrical stimulation**. If the carcass is given electric shocks immediately after slaughter, this causes muscular contractions which consume the ATP and glycogen present, leading to rapid onset of rigor mortis. The meat can then be chilled rapidly without risk of toughening due to cold shortening. The process is used for frozen lamb and sometimes for beef, to improve the tenderness of frozen carcass meat.

There is no evidence of any adverse effect on manufacturing quality. Electrical stimulation is not used for pork, whose carcasses are not generally frozen and where there is a danger of causing the PSE condition.

Mechanically recovered meat (MRM) or mechanically separated meat (MSM)

Mechanically recovered meat or MRM is the usual description of this material in the UK but it may also be known as mechanically separated meat or MSM. It is the residual meat recovered by machines from bones already more or less well trimmed by knife (see Fig. 1.5). The machines force the softer meat under pressure through perforated screens (e.g. Baader, Beehive, Bibun) or through channels formed in other ways (Protecon).

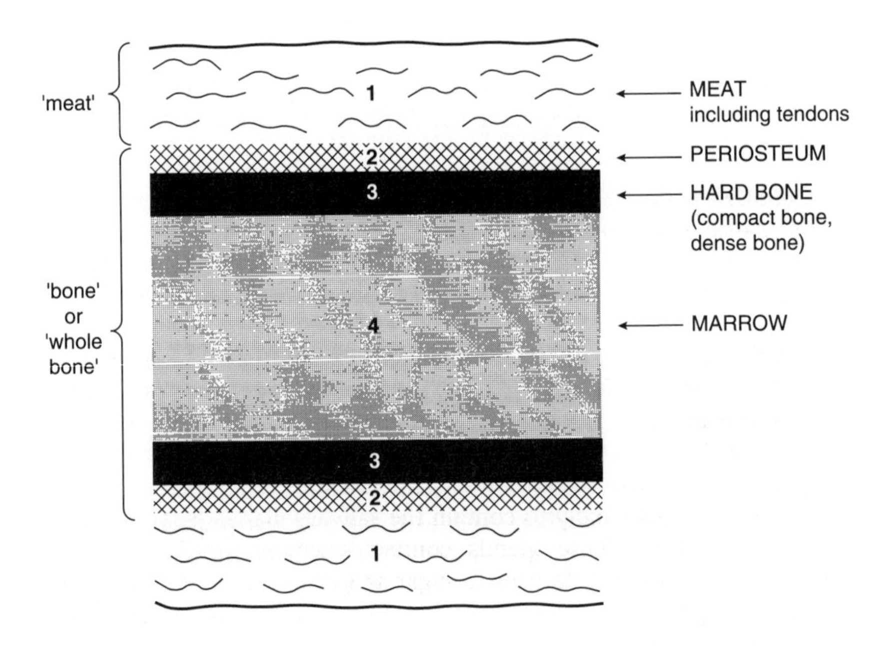

HRM may be expected to contain: 1 + some 2 only

MRM may be expected to contain: 1 + 2 + 4 + some 3

Residue from MRM may be expected to contain: 3 + some 1 + some 2 + some 4

Fig. 1.5 Diagram of a bone with adherent meat.

MRM consists of the meat and fat which were on the bones, finely subdivided by passage through the machine. Some machines also extract marrow from within the bones.

The connective tissue content is not higher than that of the most carefully hand-removed meat (HRM). Bone fragments are granular in form, not splintery, and are normally few; the actual proportion depends on the yield extracted from the machine. Current EU regulations require calcium content not greater than 0.02% and no bone fragments of size greater than 0.5 mm.

The BMMA and CLITRAVI Codes of Practice (see page 193) prohibit the production of MRM or MSM from poultry heads and feet or from the heads, feet, tails (except bovine tails) and leg bones of other animals. Following the BSE crisis in the late 1980s, regulations now also require that MRM from bovine or ovine animals:

- shall be made only on premises licensed for the purpose, and
- shall not contain any spinal cord or material from vertebral columns.

Head meat

Under normal circumstances head meat, or 'cheek meat', is a legitimate material for use in meat products, but there are certain specific problems which must be attended to:

- In the UK, under the regulations in force following the BSE crisis, meat from cattle or sheep heads is counted as 'class II specified risk material' (with a very few exceptions) and prohibited from use in the preparation of any human food; the use of pig head meat is not restricted.
- However carefully prepared, head meat is likely to be relatively highly contaminated with bacteria and should be handled and used accordingly. (The reasons include proximity of the nasal organs whose function in the live animal is to filter out harmful bacteria, thereby concentrating them, and the relatively large number of knife cuts required, which spread contamination and encourage microbial growth.)
- Head meat is also likely to contain the salivary glands from inside the animal's mouth. These glands contain enzymes (amylases) whose function is to convert starches to sugar as part of the animal's digestive process. A product made with head meat and any starchy material such as cornflour, potato or rusk may undergo rapid flavour changes due to the production of sugar, and texture may also be affected.

Fat

The word 'fat' may be used in two different senses, which must be carefully distinguished.

- In meat products technology 'fat' usually means **fatty tissue**, as opposed to lean meat or connective tissue. This is a cellular structured material whose main content is 'fat' in the other sense of the word. Examples are given in Table 1.2.

Table 1.2 Tissue fats in beef and pork

Source or location of fat	Special names		Remarks
	Beef	Pork	
Deposits on the carcass			
Subcutaneous	Body butter	Back	There is a layer of subcutaneous fat over the whole carcass. In the case of pork back, the layer is thick enough to be separated and used independently
Head		Jowl	Head fats are also sometimes separated
Deposits in the carcass			
Lining the body cavity	Peritoneal, suet		
Around:			
Intestines	Ruffle, mesenteric, gut	Flare	
Stomach	Caul, omentum		
Heart	Pericardial		
Kidneys	KKCF (kidney knob and channel fat)		
Genitals	Cod	Cod	
Trimmings			Removed during preparation of lean meat for other purposes
Intramuscular	Marbling		Considered a sign of good quality in the meat trade; probably not technically significant. Not removable by trimming. In well trimmed lean, including 3% intermuscular
Intermuscular			Intermuscular fat is contained within the muscles and is not visible

- 'Fat' also means 'chemical fat' or **lipid**. This is the main component of fatty tissue, the contents of the fatty tissue cells. The term is also used technologically to indicate **rendered fats** – lard, dripping and tallow, also vegetable oils, cooking oils, manufactured shortenings, butter and margarine. Examples of these are given in Table 1.3.

Table 1.3 Rendered fats, cooking oils, etc.

Type of fat or oil	Examples	Remarks
Rendered fat	Beef dripping	From internal fats and some body fats. '*Premier jus*' = first quality
	Block suet	Commercial shredded suet may consist of rendered beef fat, mixed with cereal (max. 17% in UK)
	Beef tallow, mutton tallow	Poorer grades from inedible raw materials (and note BSE restrictions in UK)
	Lard	From pork fat; sometimes other edible pork material. 'Greaves' is the material remaining after a rendering process
	'Free fat' (see page 34)	
Butter		Made from milk. Contains 16% water (legal limit in EU and elsewhere)
Vegetable cooking oil	Olive, cottonseed, groundnut, soya, etc.	Chemically similar to rendered animal fats but usually liquid at working temperatures
Vegetable shortening	Includes proprietary bakery fats	Made from vegetable oils by chemical hardening (hydrogenation) to make them less liquid, more plastic or solid (see page 176 for relevance to pastry). Some compound cooking fats may contain water
Margarine		Made with similar properties to vegetable shortening; may be coloured yellow. Domestic margarine made with similar properties to butter. All contain water (16% in EU and elsewhere)

Fatty tissue consists of fat or lipid, say 85%, contained in cells of connective tissue, consisting of collagen and other substances (14%) and water (11%).

Fatty tissue structure

Cells

En masse, fatty tissue cells take the usual polygonal form of biological cells. Isolated cells tend to be spherical. Size is approximately uniform, 0.095–0.15 mm.

Cells containing softer fat have thicker, stronger walls than cells containing harder fat. The differences may be considerable. For pork fats, typical observations are shown in Table 1.4.

Table 1.4 Cell walls in fatty tissue

Softness of fat	Soft	Intermediate	Hard
Fat content (%)	79.2	89.5	90.9
Cell walls			
Total %	20.8	10.5	9.1
Moisture %	15.5	8.5	7.3
Collagen %	1.0	1.0	0.7
Other solids %	14.3	1.0	1.1
Microscopic appearance	Fibrous, highly organised	Fibrous, fairly organised	Few fibres

Lipid composition and properties

Fats and oils consist chemically of mixtures of triglycerides, which in turn contain various fatty acids. Differences in hardness and softness among the fats are related to physical and chemical properties, as shown below and in Table 1.5.

Physical properties
The softer fats are more plastic at room temperature because they contain a higher proportion of fat liquid at room temperature; and have a lower average melting point or slip point.

Table 1.5 Properties of lipids

	Cottonseed oil	Pork jowl fat	Pork back fat	Pork flare fat	Beef fats
State at room temperature	Liquid	Soft	Fairly soft	Hard	Very hard
Fat liquid at 20°C (%)	100	86	85	63	0
Slip point (°C)	–	28	36	44	47–54
Iodine value	110	59	53	148	32–47
Unsaturated fatty acids (%)	72	61	54	47	47.5

Chemical properties

The glycerides of the softer fats contain a higher proportion of unsaturated fatty acids (e.g. oleic acid). That is, the structures of the fatty acids contain more double bonds or less hydrogen. They therefore have a higher iodine value (and a greater liability to rancidity).

Influence of animal feed

The fat present in an animal's diet may be used (with carbohydrate) to supply energy. Any excess over the requirement for energy purposes is normally deposited in the carcass. Excess carbohydrate is also converted into fat and deposited in the body.

In monogastric (single-stomach) animals (e.g. pigs, poultry, man):

- the composition of the lipids in the body fat tissues tends to resemble that of the fats in the diet;
- therefore, changes in fat composition of the diet will be reflected in the composition of the fatty tissue;
- in particular, pigs fed diets high in unsaturated fatty acids will themselves have generally softer body fat;
- wide variations can occur between individual animals and between groups from different farms, etc.

In polygastric (many-stomach) animals (e.g. cattle, sheep, goats):

- the first stomach contains bacteria which hydrogenate any unsaturated fats consumed, making them more saturated and harder;
- the body fat deposited in these animals is therefore harder than that of monogastrics, relatively unaffected by the composition of the lipids in the feed and relatively more uniform in composition and properties.

Hardness and softness of fatty tissue

The properties of the fatty tissue cells and the lipids within them can be summarised as follows (see also Table 1.6).

- The cell walls of softer fatty tissue have more connective tissue substances and are thicker and stronger.
- The lipids in softer fatty tissue have higher proportions of the more liquid unsaturated fatty acids.
- At chill temperature or room temperature, tissue texture is controlled by the lipid texture, and soft fatty tissue at this temperature is softer to the touch than hard fatty tissue.
- However, at body temperature, nearly all the lipid is liquid and tissue texture is controlled by the strength of the connective tissue. In this situation, soft fatty tissue is harder to the touch than hard fatty tissue.

Table 1.6 Nomenclature of hard and soft fats

Description under meat processing conditions	Soft fats	Hard fats
Physical characteristics:		
Cell walls	Tougher	More fragile
Cell contents (lipid)	More liquid	More solid
Feel of the fatty tissue:		
at chill temperature	Softer	Harder
at body temperature	Harder	Softer

In this handbook, the terms 'harder' and 'softer' are always used as they apply at **chill temperatures** or **room temperature**, i.e. under meat manufacturing conditions. This is also the usage in oils and fats chemistry. However, vets, slaughtermen and others who handle live or just-slaughtered animals use the terms the other way round, as they are experienced at **body temperatures.**

Within an animal, the softer fats are located furthest from the centre of the animal, so that

- internal body fats are hardest
- head fat is softer than back fat
- the outer layer of pork back fat is softer than the inner layer.

The animal's body temperature is lower at the outside; therefore in order to remain liquid the fat needs a lower melting point; the other properties follow.

Between animals, differences in lipid softness caused by differences in feeding (see previous section) follow the same rules and the relationship:

Softer fat = more and stronger connective tissue

appears to be universal however the differences in softness may be caused.

Between species, the relationship also appears to hold. Thus:

chicken fatty tissue	extremely soft	much connective tissue
pork fatty tissue	soft to moderate	moderate connective tissue
beef fatty tissue	hard	little connective tissue

Connective tissue

Amount

Apart from the connective tissues dispersed through the muscle and the fat (see above), note the presence of:

- sinews
- cartilage (connecting muscle to bone)
- sheaths, walls, etc, around organs and compartments in the body (e.g. diaphragm, skin).

Some typical connective tissue contents of manufacturing meats are shown in Table 1.7 below. Note the wide range of variation in each case.

Table 1.7 Connective tissue contents of some manufacturing meats

	Wet connective tissue, % of lean meat	
	Mean	Range (= mean ± 2 s.d.)
Cow beef	5.1	0.9–9.3
75/85 Lean	7.9	5.8–10.0
Neck trim	10.8	9.7–11.9
Diaphragm	8.4	3.5–13.3
Cheek	15.2	9.6–20.2
Head	14.0	8.9–19.1
Plate	12.0	6.2–17.8
Shank	13.0	5.5–20.5

Pork rind, dried rind, collagen extracts

Pork rind is edible and is commonly incorporated into meat products. Dried rind is commercially available.

Collagen extracts are made by hydrolysis of collagenous materials such as gristle, commonly but not necessarily from pork. Bone extracts are also available, made by hydrolysis of bones.

Ten per cent of the pig carcass is reckoned to be rind and only this proportion may be counted towards the 'meat content' of a product. If dried rind or collagen extracts are used, the quantity equivalent to undried connective tissue (78% moisture) should be allowed. See p. 000 for more arithmetical detail. Of course, greater quantities than these may be used, provided that the additions are declared on the product label.

Toughness

The connective tissues are formed mainly from fibres of collagen and small amounts of elastin. In young animals the collagen is partially cross-linked, flexible but relatively inelastic. With increase in age the degree of cross-linking increases, flexibility decreases and toughness increases – cf. the increase in stiffness with age in humans.

On cooking to *c.* 65°C (150°F) the collagen and elastin shrink, with some increase in rigidity and apparent toughness. Over *c.* 80°C (175°F) the collagen begins to hydrolyse to gelatin. Collagen with little cross-linking, e.g. in broiler chicken or veal, is readily softened to a gelatin jelly but the strongly cross-linked collagen of older animals is hydrolysed only with difficulty; such meat requires prolonged moist cooking for there to be appreciable softening. (Elastin is not hydrolysed at all but it makes only a small contribution to the overall toughness.)

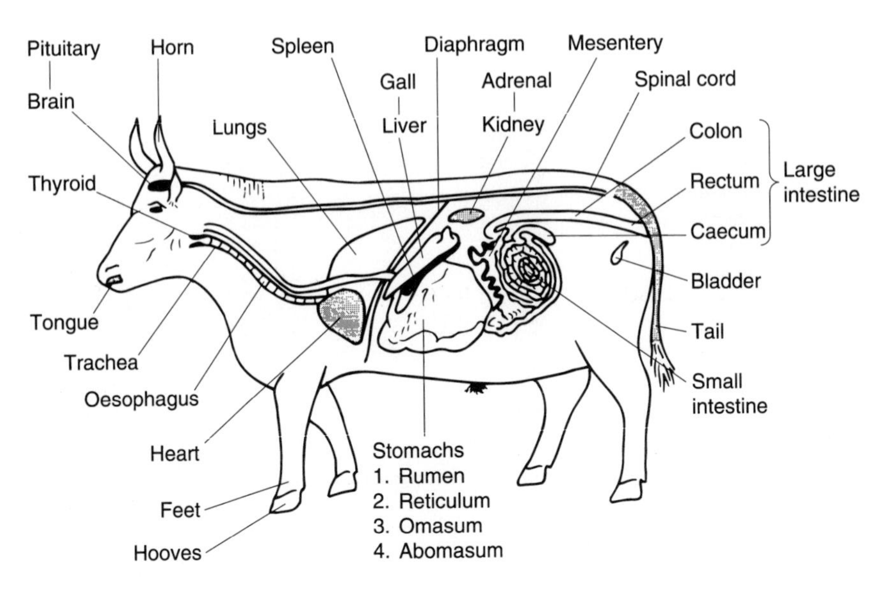

Fig. 1.6 Beef offals (from Richards, 1982).

Offals

Figure 1.6 shows the locations of the major offals in cattle. The quantities of offals available from different animals are given in Table 1.8 and some alternative common names are given in Table 1.9.

All of the offals of the main meat animals appear to have been used in domestically prepared food products at one time or another. In the UK, the

Table 1.8 Available offals (data from Richards, 1982)

	Quantities available (% of live carcass weight)					
	Cattle		Pigs		Sheep	
Contents of stomach and intestines		17.0		11.0		11.0
Skin	6.9		(10.0)[1]		17.0	
Horns[2]	0.09				1.3	
Feet	2.0		2.1		–	
		8.99		2.1		18.3
Head[2]	2.7		6.9		3.6	
Tail	0.25		–		–	
Fats	5.8		1.4		5.3	
Tongue[2]	0.65		0.4		0.3	
Diaphragm	0.27		0.4		–	
		9.67		9.1		9.2
Blood		2.2		3.0		4.1
'Red offals'						
Liver	1.3		2.9		1.0	
Heart	0.41		0.3		0.5	
Spleen[2]	0.20		0.1		0.16	
Kidneys	0.14		0.4		0.26	
		2.05		3.7		1.92
'White offals'						
Udder	1.1		–		–	
Lungs	0.64		0.8		1.0	
Stomach(s)[2]	2.34		0.5		0.5	
Intestines	1.9		2.8		3.0	
Oesophagus and trackle	0.27		0.35		0.58	
Brains[2]	0.11		0.25		0.26	
		6.36		4.7		5.34
Sexual organs	0.06		0.32		0.26	
Thymus[2]	0.05		0.16		0.19	
Other glands	0.05		0.15		0.11	
		0.16		0.63		0.57
Total		46.43		44.23		50.42

[1] Not usually counted as offal.
[2] UK regulations relating to BSE prohibit the use of these materials in human food if they derive from animals aged 6 months or over (cattle), or 12 months or over (sheep or goats).

Table 1.9 Some alternative names of offals (data from Richards, 1982)

	Pluck = Trachea + Lungs + Liver + Heart + Spleen	
Skin	Hide	(Cattle)
	Rind	(Pig)
	Pelt, fleece	(Sheep)
Feet	Cowheel	
	Trotters, toes, nails	
Head trimmings (cheeks)	Ox cheek	(Cattle only)
	Bath chaps (see page 185)	(Pig only)
Fats		
Diaphragm	Skirt	
Spleen	Melt	
Udder	Mammary	
Lungs	Lights	
Stomach	Maw	(Pig)
	Paunch	(Sheep)
Rumen	First stomach, tripe, pile, paunch	(Cattle)
Reticulum	Honeycomb	(Cattle)
Omasum	Bible, book, peck, manifold, manypiles	(Cattle)
Abomasum	Roll, reed, red, rennet	(Cattle)
Intestines, small:		
Jejunum and ileum	Runner, string, rope, rop, marm	
Intestines, large:		
Colon	After end, middle gut	
Caecum	Cop end, gut end, bung end	(Sheep)
Rectum	Bung end, fat end	(Cattle)
Oesophagus	Weasand	
Trachea	Windpipe, pipe, trackle	
Sexual organs		
Uterus	Breeding bag	
Testicles	Fries	
Penis	Pizzle	
Thymus	Heartbread, throatbread	
Pancreas	Gutbread, sweetbread	
Adrenal	Suprarenal, kidney gland	

position has been complicated by considerations arising from the BSE crisis of the late 1980s and after; see Table 1.8.

Properties

Note that:

- Only diaphragm and cheek or head meat contain muscle with similar properties to those of 'red' meat.
- Heart consists mostly of 'smooth' muscle, with properties (for manufacturing purposes) similar to but not identical with those of 'red' meat.
- Kidneys may be associated with relatively large amounts of fat (KKCF).

- The other offals have few of the functional properties of water- or fat- or meat-binding which will be discussed in the following chapters; when incorporated into meat products they may therefore may require some other binding agent to hold the product together.

2 Processing Principles

When meat is processed to make meat products there are four primary factors to be attended to. They are:

- **Moisture**. The natural moisture content of the lean meat, and any liquids added in the recipe, should be retained to a consistent optimum extent during the manufacturing process – in the interests of both yield and product quality – and through the stages of distribution, storage and any eventual cooking by the consumer.
- **Fat**. The natural fat content of the meat, and any extra fat which the product is designed to incorporate, should similarly be retained to a maximum or optimum extent throughout.
- **Connective tissue**. Where the product contains any of the tougher connective tissues, these should be presented in some more acceptable form.
- **Cohesion**. The product should retain its physical integrity.

We shall consider these in turn.

MOISTURE RETENTION

The ability to retain water is essentially a property of the **lean meat**, modified by any water and salts which may be added during processing.

Note the following definitions:

- **Water-holding capacity (WHC)**. The ability of the meat to retain the tissue water present within its structure.
- **Water-binding capacity (WBC)**. The ability of the meat to bind added water.
- **Cooking loss**. Water, fat or jelly which is lost from a piece of meat or meat mixture on cooking. In experimental work, product development or trouble shooting, it is often convenient to express the losses of water and jelly as **parts lost per 100 parts of original lean meat**, since lean meat is the most important factor in the retention of these components: on this basis the losses from mixtures of different composition may be directly compared. (Fat losses may be expressed as parts lost per 100 parts of original fat, for a similar reason.)

- **Meat binding**. The adhesion of pieces of meat to one another, especially after cooking.

Water content of meat

Lean meat

The water content of lean meat or muscle is approximately 75%, distributed as shown in Table 2.1. The forces which hold the water in the meat are not fully understood, but about 5% may be chemically bonded to proteins; 24% is held by capillary forces and may be squeezed out under pressure; and 45% is held firmly, but the mechanism is unknown.

Table 2.1 Distribution of water (%) in lean meat

	Fibres or cells		Extracellular space and connective tissues	Total
	Fibrils	Sarcoplasm		
Water	45	19	11	75
Protein	10	6	2	18
Other substances	5	–	2	7
Totals	60	25	15	100

The approximate location of the water in lean meat is shown in Fig. 2.1.

Fat

Fatty tissue contains about 10% water, located in the connective tissue of the walls of the cells which contain the lipid.

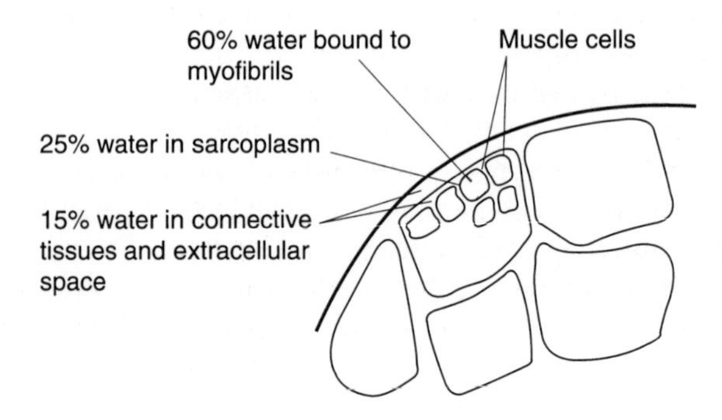

Fig. 2.1 Approximate location of water in meat.

Measurement of moisture retention

Water-holding capacity (WHC)

'Drip'

Measure by holding pieces of meat in polythene bags under standard conditions: after a standard interval the weight or volume of the liquid collected in the bottom of the bag is measured. Drip from unprocessed meat is usually quite small (*c.* 0–3%) but in exceptional cases may be higher. Large drip losses are usually associated with abnormal pH.

Pressing

Note especially the Grau–Hamm press method, commonly used in experimental work in Germany. A standard weight of sample on a filter paper is pressed between two plates, and the area of the paper wetted by liquid exuded from the sample is noted. The pressure applied is not very critical so, for example, a simple hand-operated screw can be used.

Centrifugation

The proportion of liquid removed from the meat depends on the conditions; fairly consistent results can be obtained at high centrifuge speeds.

Cooking loss

The water, jelly or fat separated from a meat product under standard processing or cooking conditions may be measured directly; the information thus obtained may be immediately relevant to both factory costs (through product yields) and likely causes of consumer complaints.

In experimental work a simplified, standardised cooking process may be used. In the system of Evans & Ranken (1975), a sample of 40 g $\pm$ 0.5 g of meat mixture is placed in a wire basket suspended by threads in a polypropylene test tube (11 cm × 4 cm). The tube is fitted with a simple air condenser and heated in a water bath at 80°C for a standard period (normally 28 min). The condenser is removed and the sample raised by the threads and allowed to cool while the cooking losses drain to the bottom of the tube. When any fat on the liquid surface has solidified, it may be punctured and the watery material below it poured off; the separate weights of 'water' loss and fat loss may then be obtained. By suitably varying the composition of the sample cooked, the effects of variations in the composition and processing of meat mixtures may be investigated.

Correlation among different measurements

Correlations are generally poor, no one measure being a good predictor of the result of any other measure. So no single test is a good predictor of the result of a manufacturing process on any particular occasion. Nevertheless, a single test, consistently applied, can be used to build up a fair *qualitative* picture of the behaviour to be expected in manufacturing practice in the long run.

Effects of pH

- Low ultimate pH, e.g. 5.2–5.5 in the back muscle (*longissimus dorsi*) results when the animal has been slaughtered humanely and quietly after adequate feeding (so there is a high ante-mortem glycogen content in the blood, therefore good post-mortem production of lactic acid). This is considered to give the best meat.
- If the pH of a single specimen of meat is changed *experimentally* by addition of acid or alkali, the minimum WHC is found near the isoelectric point of the meat proteins, about pH 5.5 or pH 4.5 in the presence of salt. This has been explained by saying that since there are fewer charged ions at the isoelectric point, attraction at that pH is maximum between the protein molecules, leaving little space for water to be bound there.
- The correlation between WHC and the *natural* ultimate pH of different meats is poor, so the ultimate pH value of an individual sample is not a good predictor of its WHC.
- The special cases of the PSE and DFD conditions are described on page 10.

Effects of added water, salt and phosphates

The separate effects of various factors can be distinguished.

Added water

The addition of water alone generally increases the yield of lean meat on cooking, despite an apparent increase in cooking loss, i.e. although much of the added water is lost on cooking, some of it is retained by the meat. Thus, in a typical case:

100 g raw lean meat ⟶ 80 g cooked meat + 20 g cooking loss (20% cooking loss)

100 g raw lean meat + 20 g water ⟶ 88 g cooked meat + 32 g cooking loss (26.7% cooking loss from the whole mixture but 8% increase in yield of cooked meat)

The maximum effect is obtained when the water is injected into the meat (as in the above example) but cooking the meat 'wet', i.e. submerged in liquid, as opposed to cooking 'dry', is also effective.

Salt

- Adds flavour (pleasant up to 2–3%).
- Restricts microbial growth (page 50).
- Interacts with lean meat proteins to give increased water retention, yield, etc; increased meat binding, cohesion, etc; increased fat binding; and texture changes.

These effects are described below.

(a) Increased water retention, yield, etc.

Cooking losses are at a minimum when the salt content, expressed as percentage of the total water in the mixture, including the water content of the lean meat, is in the range 5–8%. At these concentrations the myofibrillar proteins are dissolved and may be extracted from the meat; in extreme cases a sticky exudate is formed on the meat surface. The increases in yield and reductions in cooking loss are considered to result from structural changes within the muscle fibres as some of the protein is solubilised or extracted (with or without the formation of exudate).

(b) Increased meat binding

Protein extracted into solution forms a cement between pieces of meat, which sticks them together. In the raw state the meat becomes more sticky and cohesive; on cooking it sets to a more or less solid mass.

(c) Increased fat binding

With moderate comminution the lean meat may form a coarse network, as in (b) above, within which particles of fat are held physically. With increased comminution any free fat may be emulsified by the solubilised protein from the comminuted lean. However, in cases where much fat is present, the lean meat may not form a continuous matrix; the system may become 'fat continuous' instead and lose fat easily on cooking. See also page 42.

(d) Texture changes

Since the action of salt is on the lean meat fibres, which become more solubilised, the fibrous nature of the meat is decreased and the product becomes more gelatinous or rubbery, e.g. a frankfurter sausage, in which the binding should be so good that the sausage 'snaps', and which is completely gelatinous and non-fibrous in texture.

Note that 'salt' in all the cases above is taken to mean sodium chloride. For the other inorganic salts, see page 31.

Phosphates

The phosphates likely to be encountered are:

Orthophosphates	PO_3^{3-}
Pyrophosphates	$P^2O_7^{4-}$
Tripolyphosphates	$P_3O_{10}^{5-}$
Long chain polyphosphates e.g. 'Calgon'®	$[PO_4^{3-}]_n$ where $n = 8$–15

Only pyrophosphates and tripolyphosphates are significant in meat technology.

Phosphates alone have a salt type of action on yield, etc; 0.3% of tripolyphosphate has a similar effect to 0.7% sodium chloride. However, it is normally cheaper to use salt alone than phosphate alone.

In addition to their small direct effects on water binding, pyrophosphates and tripolyphosphates act **in the presence of salt** to increase greatly the effect of the salt on lean meat proteins (see above), i.e. they increase the effect of salt on cooking loss, yields etc., on meat binding, fat binding and texture, *or* they act catalytically, accelerating the salt effects and allowing the same results to be obtained in shorter time.

The combination of these phosphates with salt is most effective in lightly heated meats (e.g. pasteurised hams, luncheon meats), less effective in sterilised products.

Phosphate flavour is bitter and is usually considered unpleasant at 0.3–0.5%.

Effects of water, salt and phosphates together

Table 2.2 shows the cooking losses and the corresponding yields found in laboratory experiments in which various mixtures of lean meat, water, salt and phosphate were processed in three different ways.

Modes of action of salt and phosphates

Salt (sodium chloride)

The action of salt is usually spoken of as a 'solubilising' or 'hydrating' action on the myofibrillar proteins actin and myosin. (The lowest cooking losses

Table 2.2 Cooking losses and yields of various meat mixtures (data from Ranken, 1984)

Mixture no.	Mixture containing:	Unheated (held 24 h at 5°C)	Pasteurised (ham process)	Sterilised ($F_0 = 3$)
1	100 Meat only	Lost 0 Yield 100	Lost *c.* 20 Yield *c.* 80	Lost *c.* 30 Yield *c.* 70
2	100 Meat + 20 Water	Lost 4–10 Yield 110–116	Lost *c.* 28 Yield *c.* 90	Lost 35–45 Yield 75–85
3	100 Meat + 20 Water + 2 Salt	Lost 0 Yield 122	Lost 20–25 Yield 97–102	Lost 28–45 Yield 77–96
4	100 Meat + 20 Water + 0.3 Phosphate	Lost 2–6 Yield 14–118	Lost 22–28 Yield 92–98	Lost 30–45 Yield 75–90
5	100 Meat + 20 Water + 2 Salt + 0.3 Phosphate	Lost 0 Yield 122	Lost 12–20 Yield 102–110	Lost 25–40 Yield 82–97

Notes

(1) Mixtures made with 100 parts of diced pork meat, injected with the quantities of water, salt and sodium tripolyphosphate in the quantities shown. Drip and cooking losses and yields measured under laboratory conditions; the values given are approximate. All tests repeated three times.

(2) The range of variation in most of the experiments (three results each) is quite high. This is probably a result of slight unevenness in composition among the individual meat dice in each experiment. Variations may be expected under manufacturing conditions, for a similar reason.

(3) The incorporation of water alone increases yields in every case.

(4) The addition of salt, without phosphate, produces further increases in yield in every case.

(5) The addition of phosphate without salt also produces increases in yield, smaller than those produced by salt without phosphate.

(6) Salt and phosphate together produce increases in yield, greater than the combined effects of salt and phosphate individually.

(7) As would be expected, losses are greater and yields lower when the mixtures are pasteurised, and more so when they are sterilised. The effects of salt and phosphate are also much smaller in the sterilised samples.

are obtained when the concentration of salt in the water, including the water present in the meat, is in the range 5–8%. These are also the concentrations at which solutions of myofibrils can be extracted experimentally from lean meat.)

Other inorganic salts

All inorganic salts increase the WHC of meat in the same way. This property is exhibited in proportion to the ionic strength of the salt in solution. (Ionic strength is a function of the size and the electric charge of the ions composing the salt. Its precise definition need not concern us here.)

Sodium citrate and potassium chloride are sometimes used as ‘cutting aids', in addition to common salt, to increase the salt effect without excessive increase in salty flavour. Most other inorganic salts are either poisonous or too expensive for such use.

Phosphates

- In the absence of salt the phosphates have only their own 'ionic strength' effect, which is quite small (see Table 2.1).
- They act synergistically or catalytically with salt. Small concentrations, 0.3% or less, are sufficient if properly distributed; higher concentrations are sometimes used in practice to allow for non-uniformity of distribution.
- The effective phosphates are pyrophosphate, tripolyphosphate and some higher phosphates. They apparently depend for their effect on hydrolysis to diphosphate (pyrophosphate) by phosphatases in the meat. All phosphates are hydrolysed by meat enzymes to orthophosphate in 24–48 h. Orthophosphate itself is not an effective catalyst of the salt effect but the effect of the higher phosphates remains even after they have been converted to orthophosphate. The action may be a triggering mechanism, possibly linked with the hydrolysis step pyrophosphate ⟶ orthophosphate, but the full explanation is not clear.
- Note the practical point that phosphates are soluble only with difficulty in salt solutions but dissolve more easily in plain water; mixed solutions should therefore be made by dissolving the phosphate first.

'Hot' meat processing

Meat comminuted with salt immediately post mortem retains a very high WHC. This appears to be due to the adenosine triphosphate (ATP) which is still present in the meat at that time. The action appears to be similar to that of inorganic polyphosphates.

Penetration and distribution of water and salts

Since the effect of salt appears to be primarily a result of chemical interaction with the meat protein (hydration or solubilisation), it is obvious that the maximum effect is obtained only when the salt becomes uniformly distributed through the fine structure of the meat and in close contact with the protein.. The same is true for added water, phosphate and other salts.

Optimum yield occurs when the salt concentration is uniform throughout the lean meat and, ideally, 5–8% of the water content.

Various factors affect penetration:

- **Comminution**. The smaller the meat particles, the easier and quicker the penetration of other substances.
- **Immersion or injection**. If large pieces of meat, e.g. hams, are cured by immersion in a brine, sufficient time must be allowed for salt and water to penetrate before the product is cooked.
- **Time**. Holding untreated lean meat, or meat with moderate amounts of added water, for up to 3 days before cooking has a negligible effect on the cooked yield, but with added water in the presence of salt or salt plus phosphate, yields can be increased by 3–4% compared with the same mixtures processed without delay. In modern processes where brine is injected into the meat before immersion to speed up the curing process, time is still necessary to allow penetration from the sites of injection into the meat which lies between the sites.
- **Mechanical action**. This increases penetration by a combination of rubbing the meat surfaces and flexing the meat pieces.
    - **Tumbling** or **massaging** are the processes most commonly used. Tumbling or massaging for short periods, especially with mixtures containing salt or salt plus phosphate, produces greater increases in cooked yield than holding similar mixtures for several days without movement. See page 130 for the machines used. Tumbling or massaging also produces an exudate from the meat which is important for meat binding. This is considered in more detail under that heading (page 44).
    - **Other forms of mechanical action**, mixing, flow in pipelines, through machines, etc.; mincing, chopping, etc. (in addition to the cutting action). Squeezing, pressing, etc. of the meat occurs in all of these, with similar results to tumbling or massaging.

Effects of fat on moisture retention

(a) The 'marbling' fat present in otherwise lean meat (around 3–10%), also any free fat in a meat mixture, can have a small direct effect in reducing the moisture losses from the lean meat. The reduction however is small and may usually be ignored in practice. ('Lean meat' in the trade is taken as 90–100% 'VL' or 'visual lean'. Meat as used for manufacturing usually contains 'visual' fat in addition, e.g. '85% VL' with about 15% fat content or 'semi-lean meat' or '50% VL', with about 50% fat content; these proportions may or may not be closely specified.)

(b) A more important effect is partly arithmetical and arises if the fat content of the meat used in a recipe is allowed to vary. If the fat content of the meat is increased, with no other change:

- the lean content is decreased
- therefore the water present in the lean is decreased
- therefore the total water in the recipe is decreased
- therefore the salt content of the water is increased
- therefore water loss on cooking may be decreased and yield increased
    - because there is less total water in the mixture to start with
    - because of the increased concentration of salt-on-water
    - in addition there is a slight enhancement of the salt-on-water effect due to the presence of fat, as in (a) above.

but, of course, the fat loss may be increased because of the higher fat content.

FAT RETENTION

Cooking losses from fatty tissue

If a joint of meat is roasted or a slice of bacon fried, it is obvious when cooking is complete that the majority of the original fat is still in its original place. The fat is retained inside the fatty tissue structure, even though it has become completely liquid during the cooking and might be expected to have drained away if that were possible.

For fat loss to occur from a meat product, the first requirement is that at some stage the fatty tissue cells are broken and the fat within them released. The fat after release is called **free fat** to distinguish it from fat held within fatty tissue.

To a large extent, the factors controlling fat loss operate independently of those controlling water loss. The conditions under which fatty tissue cells are broken are the major consideration; other factors are secondary or even unimportant.

Effects of comminution

Cutting, dicing, slicing, etc.

If fat is cleanly cut into pieces, the weight of fat lost on cooking agrees closely with the calculated value for the number of cells cut through on making the pieces. That is, the number of cut cells is the only significant factor controlling the cooking loss. The effect is independent of the softness of the fatty tissue, since the cells in all kinds of fatty tissue are of about the same size. For 1 cm dice ($\frac{1}{2}$in), the loss is about 6%.

Mincing

Mincing leads to very variable cooking losses, depending on several factors:

(a) Softness of the fat. **The harder the fat the greater the loss**. Typical figures are given in Table 2.3.

Table 2.3 Fat cooking losses from minced pork fat (data from Ranken, 1984)

	Cooking losses
Jowl fat	20–26%
Back fat, soft	29–38%
Back fat, hard	49–73%
Flare fat	85–86%

The causes of the variations include:

- the stronger cell walls in soft fat are more resistant to damage while passing through the mincer;
- the softer cell contents make the cells more pliable and better able to avoid damage.

(b) Design, setting and condition of the mincer. This may affect the result, but not very predictably, in the following ways:

- larger apertures and clearances, including those enlarged by wear, will permit more material to pass undamaged by the mincer surfaces;
- failure to cut other connective tissues cleanly may result in longer residence times, greater turbulence and greater risk of damage.

(c) Temperature of mincing (see below).
(d) Presence of other material (water, meat, etc.) in the mincer (see below).

Chopping

Chopping in a bowl chopper produces a large number of clean cuts in the material, together with vigorous stirring.

Chopping fatty tissue alone, previously diced or in large pieces
This gives losses related to:

- Degree of comminution (i.e. length of time in the chopper, affected to some degree by the sharpness, setting and shape of the blades).

- Softness of the fat. The effect here is less marked than with mincing; it is notable after short chopping times (i.e. with coarse chopping) but the differences may disappear with longer chopping.
- Some typical figures are given in Table 2.4.

Table 2.4 Fat cooking losses from chopped fats (data from Ranken, 1984)

		Fat loss, g/100 g fatty tissue, chopped dry for:						
		10 s	30 s	1 min	2 min	3 min	4 min	5 min
Soft pork								
Back fat	No. 1	6	16	29	44	47	49	53
	No. 2	–	–	27	39	49	57	69
Hard pork								
Back fat		11	24	43	60	69	73	75
Beef clod fat	No. 1	–	42	52	59	71	79	84
	No. 2	–	44	54	64	74	81	83

Chopping fatty tissue alone, after previous mincing
This has the following effects:

- Damage caused by chopping starts at the level already caused by mincing.
- With hard fats, already greatly damaged by mincing, chopping causes little additional cooking loss.
- With softer fats, less damaged by mincing, short chopping times (under 1 min) may still give only moderate cooking loss, but the chopping time to give maximum loss (in the region of 80%) is shorter than with previously unminced fat.

Freezing before comminution

- Freezing followed by complete thawing before comminution: cooking losses unaffected or only marginally increased (see Table 2.5).
- Freezing followed by mincing in the frozen state: cooking losses of all fats very high.
- Freezing followed by chopping in the frozen state: greater cooking losses than by mincing in the frozen state.

Chilling before comminution

At temperatures above the freezing point of water, the loss on mincing is lower the higher the temperature. Some typical figures are:

Table 2.5 Effects of freezing and comminution on fat cooking loss: cooking losses as per cent of fatty tissue. (Data from Ranken, 1984. Results on the same horizontal line refer to the same sample of fatty tissue.)

Pork fatty tissue	Minced unfrozen	Minced thawed	Minced frozen	Chopped unfrozen	Chopped thawed	Chopped frozen
Jowl	11.4	9.7	71.2	19.6	22.3	77.4
Back	21.3	20.6	77.4	45.1	47.3	83.1
Flare	75.9	73.2	75.0	82.6	85.0	82.4

Notes:

- In frozen fat it is the water content of the connective tissue which is frozen, making the cells completely rigid (the actual frozen temperature, in the range −5 to −25°C (+23 to −13°F), makes no difference).
- Mincing in this state causes almost complete destruction of the cells and release of the lipid fat.
- Note that for the hard flare fat the losses were high when minced in the unfrozen state, and mincing when frozen made no further difference.

Temperature of mincing	3°C (37°F)	5°C (41°F)	17°C (63°F)
Fat loss on cooking	58%	48%	38%

With chopping, the effect is similar but smaller (i.e. lower losses at higher chopping temperature).

The explanation here is that while the water content of the cell walls is not frozen at any of these temperatures (therefore the cell walls remain flexible), the lipid in the cells is more solid at the lower temperatures, which makes the cells relatively more rigid.

Comminution in the presence of water

In the presence of moderate amounts of water pork fatty tissue can be chopped with little or no increase in subsequent fat loss on cooking (Fig. 2.2). This is probably because

- the water softens the connective tissue between fat cells, and
- the more fluid mixture allows fat cells to avoid the cutting edge of the knives;
- the combined effect of these factors is to tease out individual fat cells or clumps of cells instead of cutting through them (see Fig. 2.2).

Damage to the fat cells is reduced in the same way when pork fat is chopped in the presence of other wet, fluid material, e.g. comminuted lean meat.

In the case of **beef fats** however, chopping in the presence of water causes a small increase in cooking loss. This may be because the cell walls are so thin that any softening by the water makes them not more pliable, as with pork, but more fragile.

Fig. 2.2 Specimens of pork back fat retained on a filter paper marked in 1 cm squares. (a) Chopped 4 min without added water. (b) Chopped 4 min with added water, 50 g per 100 g fat (from Ranken, 1984).

Effects of heating

Heat damage to cells

Under the microscope, individual fatty tissue cells can be observed to break and lose their contents as the temperature is raised through the range 40–80°C (104–176°F). This does not appear to happen to a great extent, however, in meat products.

Melting of fat

Melting of fat occurs at about 35–40°C (95–104°F), depending to a certain extent on the type of fat. When the fat is molten it is able to escape from the fatty tissue cells if these have been damaged, becoming 'free fat'. This can escape from a meat product unless the product has been made with a structure capable of retaining it.

Binding of free fat

Pre-formed emulsions

Pre-formed emulsions are sometimes made as intermediate products in the formulation of sausages, burgers, etc. Their function is to convert excess pork fat, trimmings, etc., into a stable form which can be minced or chopped and incorporated into other products without leading to significant fat losses on heating.

Pre-formed emulsions, soya-based

The stability of soya-based pre-formed emulsions is governed by the following factors:

- Fat which remains in its fatty tissue cells does not require to be further stabilised. In formulation, therefore, the fat to be stabilised (emulsified) is only the free fat liberated when the fatty tissue is comminuted. Under typical recipe conditions this is about 50% of the fatty tissue.
- The free fat is held in a soya–water matrix in which the ratio of protein to water determines the proportion of fat which can be held in a heat-stable form. A protein–water ratio which is found satisfactory for a certain amount of free fat may be unsatisfactory if the proportion of free fat is increased.
- Note that differences in comminution procedure, temperatures, etc. (e.g. use of frozen fat) may alter the proportion of free fat for the same amount of fatty tissue.

For a typical soya isolate the compositions of heat-stable mixtures are given in region A of Fig. 2.3. Other soya products give diagrams of similar form.

A typical formulation is:

Pork back fat	50 parts
Water	50 parts
Protein, e.g. soya isolate	6–10 parts

This should be chopped at high speed until a fine cream is formed. For soya isolate cold water is usually best. The protein–water matrix sets on heating to form a solid mass which traps the free fat.

This is not a true emulsion and all its properties are not yet clearly understood.

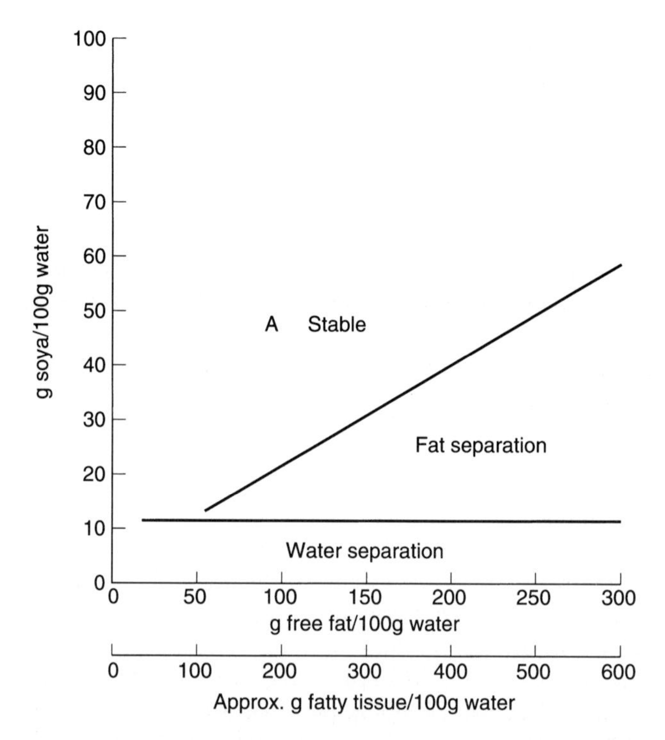

Fig. 2.3 Stability of water–fat–soya emulsions (from Ranken, 1984)

Pre-formed emulsions, caseinate-based

Caseinate-based emulsions appear to be more like true emulsions than the soya products. In the cold, approximately 20% caseinate on the water content is satisfactory for up to equal proportions of water and fat, i.e. a '5 : 5 : 1' formula, water : fat : caseinate. With hot water (95°C, 203°F) the proportion of caseinate may be reduced to 0.6, as in a soya-based formulation.

Pre-formed emulsions, rind-based

Pre-formed emulsions may also be made using pork rind as the emulsifying agent. The rinds are pre-cooked, usually along with the fat, for 1 to $1\frac{1}{2}$ hours in hot or boiling water and used hot. The inclusion of a small proportion of soya isolate or sodium caseinate gives a firmer emulsion. Typical proportions are:

Pork back fat*	7*	7*	14
De-fatted rind*	7*	7*	7
Soya isolate		1	
Caseinate			1
Water	7	7	14

* Or 14 parts of fatty rind (50% fat).

If the emulsion is not to be used immediately, it must be cooled rapidly and kept cold to avoid spoilage or loss of the firmness due to the soya component.

Emulsions may also be made on similar principles, using chicken skin in place of pork rind.

Effect of salt

Protein–fat emulsions do not normally keep very well, sometimes due to imperfect care of the ingredients before manufacture and sometimes because some heating occurs during production, with difficulty in cooling down sufficiently afterwards. Salt improves the keeping quality (about 2% salt should be used).

However, salt interferes with formation of the emulsion, so it should be added at the end of the chopping process, when the emulsion is already formed. Allow a final 30 seconds mixing to incorporate the salt uniformly.

Meat 'emulsions'

(These mixtures are often referred to as 'brat' in English though this is not the proper meaning of the German word, which means 'roast' or 'grill'. The French use the term 'pâte' or paste.)

In these mixtures any free fat must be bound by lean meat so that it will not be lost on cooking. The lean meat must be made to form a **matrix** to hold the fat; at the same time the lean and the matrix material are required to bind the product together, to bind any added water and to provide meaty texture.

However, there are conflicting requirements. The conditions which produce the best binding effects (e.g. intense chopping) are also the conditions which increase the amount of free fat requiring to be bound in the mixture.

Methods of dealing with these conflicting requirements, for coarse-cut and for fine-chopped products, are discussed further on page 132.

In the case of coarse- to moderately fine-chopped products, the technology is designed for maximum retention of fatty tissue cellular structure. To minimise the production of free fat from the fatty tissue, the following points should be noted:

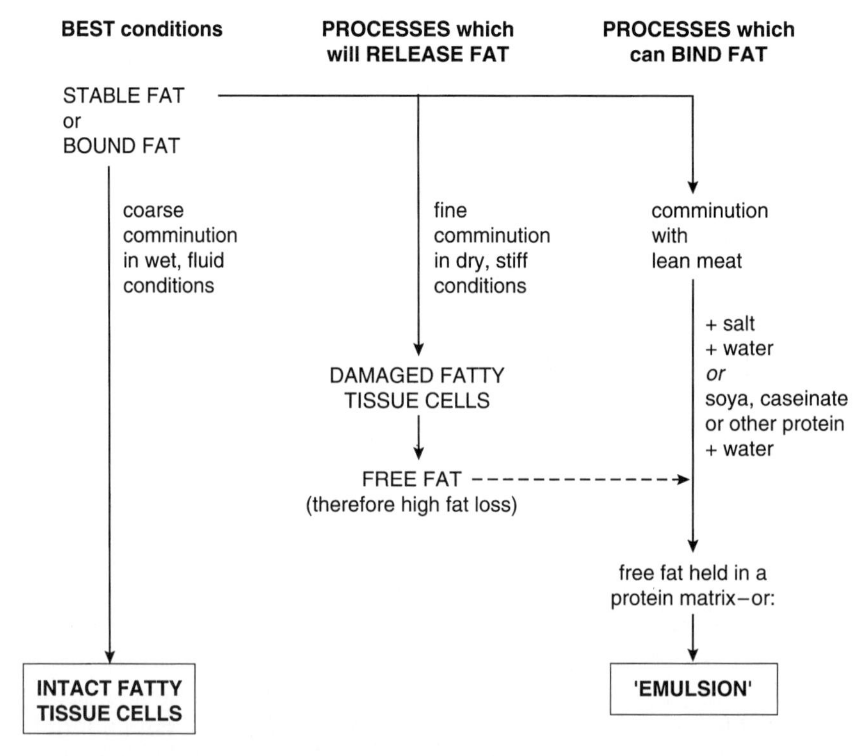

Fig. 2.4 Fat binding in meat products – summary.

- soft fats should be used
- mixtures should not be frozen
- mincing should be avoided
- pork fat should be comminuted with water or low-viscosity meat mixture
- fat should be comminuted as late in the process as possible.

In the case of *fine-chopped products* ('emulsion products'), there is considerable production of free fat during the chopping but the free fat should be retained by the lean meat–water–salt matrix which is produced at the same time.

These features are summarised in Fig. 2.4.

Fat in pastes and pâtés

The fat in the body of a pâté or paste is mainly or all in the free form. It is not firmly held in a lean meat matrix but is apparently adsorbed on to lean meat particles. There it acts as a plastic lubricant between the cooked meat particles, thus giving the product its spreadability.

There is a practical maximum fat content (about 15%) which can be held in this way. At higher levels fat separation occurs when the product is cooked. In some pastes and pâtés this is deliberately accepted and the extra fat is allowed to form a layer on top.

Other questions

Practical problems related to softness of fatty tissue are dealt with elsewhere (e.g. slicing of bacon, page 157; rancidity, page 94).

CONNECTIVE TISSUE

Tough connective tissues, difficult and unpleasant to chew, are found in

- meat from older animals, e.g. cow beef;
- meats such as beef flank or diaphragm where there were no bones to provide support in the live animal.

These connective tissues can be softened and rendered more acceptable to the taste by prolonged moist cooking. Domestic recipes for stews or casseroles are designed to ensure this. Some manufacturing processes, e.g. for canned meats, can achieve it also.

However, many meat products are not intended to be cooked in such ways but still they may need to be made with meats containing tough connective tissue. Here the toughness problem must be dealt with by comminution of the connective tissues to produce small fragments which will need less chewing.

It may be adequate to mince the meat, but unless the mincer blades and plates are sharp the operation may be difficult, with blocking of the mincer. Chopping in a bowl chopper (also with sharp blades) or mill gives better results.

MEAT BINDING

Some binding occurs when cut surfaces of lean meat come into close contact. This need not be under pressure but the binding is enhanced if pressure is also applied. The binding is stronger again if:

- there is meat 'exudate' on one or both surfaces (4–9% of the total meat appears to be ideal);
- the mixture is heated to 65°C (149°F) or higher.

Binding between or on to intact connective tissues such as muscle sheaths or skin (e.g. pork rind) is rather poor, but connective tissues, free fat or pieces of fatty tissue can be included in a mixture already bound in a matrix by meat exudate.

Measurement

The extent and strength of binding are most conveniently assessed subjectively by cutting thin slices from the cooked product and observing whether or not the slices fall apart either spontaneously or on exertion of a small force.

Exudate formation

Massaging and tumbling

These are the commonest forms of mechanical action used in modern technology for the formation of exudate. They are essentially similar but massaging is more gentle than tumbling. (See page 150 for descriptions of machines in use. Exudate may be formed by other forms of mechanical action also, see page 33.)

The exudate consists of a solution of sarcoplasm and myofibrillar proteins with 5–8% salt.

Behaviour of different meats

- 'Softer' meats produce exudate more easily under comparable conditions. The practical sequence is probably: pork and chicken (produce exudate most easily) – turkey – mutton – beef (produces exudate least easily).
- Exudate is not produced at muscle surfaces where the connective tissue is intact. Cutting or scoring the meat therefore assists exudate formation and binding.
- Among different samples of the same species or even of the same cut of meat there may be wide variations in capacity to form exudate, from copious formation to almost none. Usually in manufacturing conditions these will be mixed together.

Relationship between exudate formation and product texture

- Meat binding increases the sliceability of the cooked product; integrity of thin slices is a good practical test of the effectiveness of binding (page 43).
- Since the exudate consists of a solution of the fibrous meat proteins, it is formed at the expense of fibrousness of texture in the final product.
- On heating, the exudate sets to a solid gel, so: more exudate ⟶ texture more gel-like, less fibrous.

- With excessive formation of exudate the final product texture may be quite rubbery.

Meat exudate and fat binding

- The presence of exudate can cement fatty tissue into a meat mixture.
- Free fat or oil can also be incorporated into a system containing exudate. It is not clear whether this fat is held in a true emulsion, but probably it is not.

Binding at low temperature

Some re-formed meat products are made by forming under high pressures at carefully controlled temperatures of −2°C (28°F) or below.

Flaked meat is usually used, with or without the addition of salt. The mechanism of binding under these conditions is not clear.

It is not likely that solubilisation and extraction of proteins could occur until ice in the starting materials melts to form water. This may occur due to heat produced by the mechanical work of processing. In high-pressure systems the pressure may be great enough to melt ice momentarily. Damage to cells and connective tissue may also occur under high pressure, which may assist breakdown and extraction of protein, therefore retaining water and increasing binding.

Similarities and differences among meats

Differences in response to tumbling, exudate formation and binding between species and among samples of the same kind of meat (even between different muscles) are referred to above (page 44). Quite wide differences in cooking loss of the unprocessed meat, and in the size of the response to added water, salt, etc., also occur, even among samples of the same kind of meat.

These differences occur unpredictably and their causes are not understood at present. In practice they are usually 'averaged out' when moderate- or large-sized batches of product are made.

An arithmetical influence of fat on the salt–water–lean meat effect is discussed above (page 33), where increase in the fat content of a product may increase the salt concentration in the lean meat. A similar effect may occur if there are wide variations in connective tissue content of the meat in use. In theory, if the relevant fat and connective tissue contents were known, these factors could be compensated for by adjustment of the salt content of the recipes.

Within the range of variability imposed by the above factors, there is no evidence of significant differences in behaviour between

- meat from different commercial cuts, including the 'noble' cuts not normally used for manufacture;
- meat from different grades of carcass;
- meat of different species.

BINDING AIDS

A number of materials which gelatinise or set on heating may be used as binding aids, especially in cheaper products with low proportions of lean meat. They are used:

- to bind free fat and reduce fat cooking losses
- to improve cohesiveness of the product.

The aids include:

- blood plasma
- fibrin + fibrinogen mixtures, e.g. 'Fibrimex'
- extracted bone or gristle preparations, e.g. 'Collipro'
- soya isolates (soya protein with about 12% moisture)
- soya concentrates (de-fatted soya, about 12% moisture, 53% protein, 32% carbohydrate)
- milk powder
- whey powder
- casein and caseinates
- starches
- carrageenans
- egg white.

In choosing a binding aid, the following factors should be taken into account:

- Proportion technologically necessary – suppliers will normally give advice but this should be tested in trial batches; see also pages 39–40 on the use of soya isolate and caseinate.
- Effects of that proportion on the appearance texture and flavour of the product.
- Cost.
- Label declarations required: product name and description, ingredient list, meat content; note particularly that almost none of the binding agents listed above may be counted as contributing to the meat content.

PROCESSING LEAN, FAT AND CONNECTIVE TISSUE TOGETHER

When making any given product the principles outlined above are balanced to produce the optimum effects for that particular product. In most manufacturing operations one or both of the processes shown in Fig. 2.5 are brought into action.

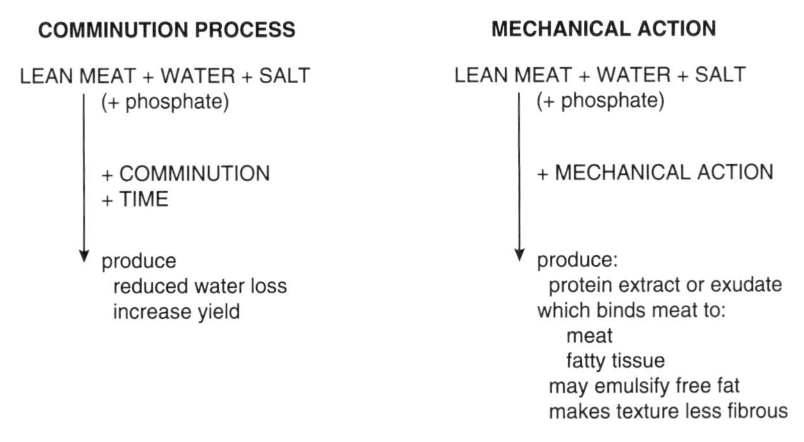

Fig. 2.5 Comminution and mechanical action – diagrammatic summary.

Although for maximum water binding fine comminution of the lean meat may be desirable, comminution may be restricted for various reasons:

- an obviously coarse-cut product (hamburger, some sausages) may be required;
- a more fibrous, less gelatinous or rubbery texture may be required;
- there may be problems of fat comminution (see page 34).

Products such as hamburger and coarse-cut sausage therefore require compromises:

- sufficient binding (exudate) to hold the product together may be at the cost of increasing gelatinous or rubbery texture;
- retention of maximum fibrousness in the lean meat fragments may be at the cost of poorer binding or integrity of product.

Figure 2.6 illustrates some of these factors.

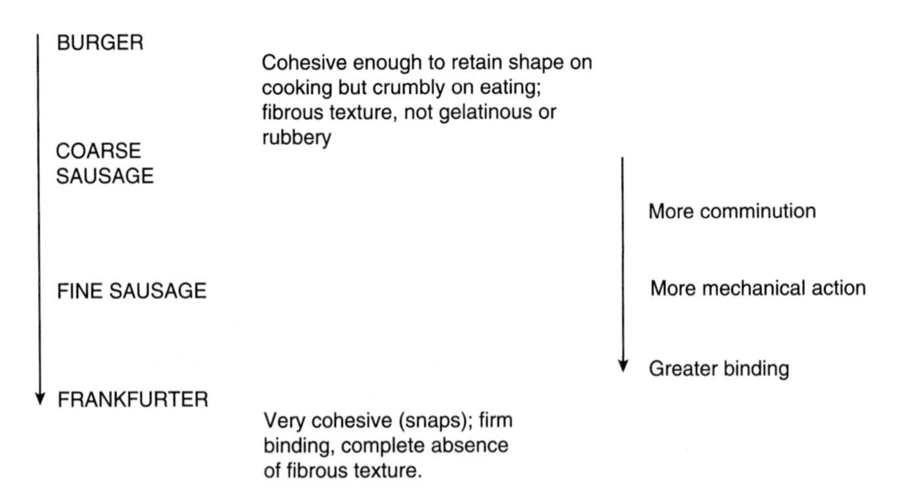

Fig. 2.6 Degree of comminution of different meat products.

Note also the effect of process variables such as delay and re-working, which tend to move product texture downward in the series shown in Fig. 2.6.

3 Curing

Preservation by curing with salt is very ancient; it was used by the Egyptians *c.* 2000 BC. In its original meaning 'curing' meant 'saving' or 'preserving'. Food-curing processes therefore include preservation processes such as drying, salting and smoking. For meat products, the term 'cured' is now usually taken to mean 'preserved with salt and nitrite'. The terms 'curing chemistry' and 'curing reactions' usually refer to the formation of cured meat colour by the action of nitrite.

GENERAL PRESERVATION PRINCIPLES

Meat is susceptible to bacterial decomposition, which results in the production of 'off' odours, followed by slime production and structural breakdown. The aim of curing is to prevent or delay this natural process of decomposition. This is done by changing the properties of the meat so as to prevent the growth of the bacteria which would otherwise cause rapid decomposition. The meat composition is altered mainly by use of salt (sodium chloride), sodium nitrate, and sodium nitrite. Additions of sugar, spices, etc. also have significant although smaller effects.

As the conditions become less suitable for the spoilage bacteria, they become more suitable for other species of bacteria which can tolerate salt, nitrite, nitrate, etc. The net microbiological process which occurs during curing is therefore the **replacemen**t of the **meat flora** (mainly pseudomonads) by a **cured meat flora** (mainly lactobacilli and micrococci). The bacon flora will still spoil the meat but less rapidly; most bacon or ham will not keep indefinitely.

The engineering of these changes, i.e. the practice of curing, requires care. The unwanted 'meat' bacteria should be present in the minimum numbers at the beginning of curing; then conditions must be maintained which will permit optimum development of the cured meat bacteria. Points to observe include the following.

Pre-slaughter

Rested animals have fewer organisms in their tissues than animals which have been stressed by travelling, starvation, poor lairage conditions, fighting, etc. Exhaustion before slaughter may result also in high pH post mortem, which will encourage microbial growth (page 10).

During slaughter

It is essential to keep the level of bacterial contamination on the meat as low as possible, by proper sanitation and hygiene during slaughter and evisceration.

Post-slaughter

Bacterial numbers on the carcass will inevitably increase but the degree of increase should be kept to a minimum by:

- rapid reduction in the temperature of the carcass (e.g. for pigs, to 3–5°C (37–41°F) within 16 hours after slaughter);
- control of humidity: the atmosphere should be kept reasonably dry;
- short holding time of meat before curing (*c.* 24 hours).

EFFECTS OF THE MAIN CURING INGREDIENTS

Salt

Preservative action

'Salt inhibits objectionable putrefaction and dangerous micro-organisms, and those which it does not inhibit are more or less unobjectionable' (Ingram & Kitchell, 1967). The effect is mainly due to salt concentration in the water in the product (sometimes called 'brine concentration').

A level of 4% salt-on-water with no nitrite or other additives will preserve sliced vacuum-packed bacon for 3 weeks at 5°C. The level of 3.5% salt-on-water is taken as the lowest level for safety in shelf-stable canned cured meats.

Note: 4% salt-on-water = 3% salt-on-lean meat (lean meat with 75% water)
= roughly 1.5% salt on fatty meat with 50% fat content

Selective microbiological effects in brines

In brines of 22–25% salt (i.e. saturated with salt or close to saturation), the bacteria which convert nitrates to nitrites are not inhibited. Such a brine, containing these organisms, is a 'live' or 'active' brine as used, for example, in Wiltshire curing (page 154).

Flavour

About 3.5–5% salt in the product is probably about the present-day upper limit of acceptability, depending on the product. About 1.5–2% is a usual average acceptable level; note that this is just compatible with the concentrations required for preservation (above).

Water binding and meat binding

The effects of salt on water and meat binding, particularly when enhanced by mechanical action, are of special value in the production of some kinds of ham or combination cured meats.

Water content and water activity

Since the preservative action of salt depends on its concentration in the water, increased preservation can also be achieved by reducing the water content, i.e. by drying. The salt/water relationship can be expressed as the water activity (a_w) of the system. Some significant water activities are shown in Table 3.1.

Table 3.1 Water activities of meat and meat products

	a_w
Pure water	1.0
Raw meat	about 0.99 (0.98–0.99)
Uncooked cured meats intended to have long shelf-life without refrigeration, e.g. dry sausage, traditional raw ham	about 0.92 (0.70–0.96)
Cooked cured meats intended to have long shelf-life without refrigeration, e.g. some cooked ham	about 0.97 (0.96–0.98)

There are simple instruments (e.g. Lufft) for measurement of the a_w of meat samples but the measurements are subject to large errors – at least $\pm$ 0.02 – and it is difficult to obtain reproducible results.

Nitrite

Nitrite salts, normally acting in the form of undissociated nitrous acid (HNO_2), are powerful preservatives against all spoilage and food poisoning organisms: they are the basis of all traditional and modern cured meats. Unfortunately, however, in excess they are poisonous to humans. For that reason the quantities permitted in foods have been restricted for as long as there have been preservatives regulations. In the 1960s and 1970s they were found to be implicated also, under certain conditions, in the formation of carcinogenic nitrosamines. There was intense investigation of the quantities necessary to provide safety from food-poisoning bacteria, including *Clostridium botulinum* spores in particular, with minimum risk of nitrosamine formation. The present regulations in all countries are framed accordingly, as also are all the recommendations which follow in this handbook.

Preservative action

Uncooked cured meat products

Examples of these are unpasteurised, unsmoked and cold-smoked bacon. In these cases, the preservative effect is due mainly to the residual nitrite content in the product. This diminishes with time, at a rate which depends on the storage temperature. When the residual nitrite content approaches zero, spoilage is usually observed then or shortly afterwards, as seen in Table 3.2.

Usually, the total aerobic count of the bacon will reach 10^7 organisms per gram, either at about the same time as the appearance of marked off-odour (bacon with low nitrite content) or some time before off-odour (bacon with high nitrite content). On reaching 10^7/g, counts usually increase only slightly (to 10^8/g) or not at all. The limit of 10^7/g therefore provides a somewhat conservative estimate of the shelf-life of vacuum-packed bacon.

It is difficult to set a numerical value to the minimum nitrite content required to give a commercial shelf-life because of the many variable factors involved, some of which may not be known in every case, e.g. salt content, initial microbial contamination, storage temperature.

One useful attempt can be summarised as follows. If commercial shelf-life is defined as stable in vacuum pack for three weeks at 5°C, 41°F (i.e. with total microbial count below 10^7/g), *then* bacon with satisfactory commercial shelf-life is given by the following combinations:

Salt-on-water (%)	Initial nitrite content (ppm $NaNO_2$)
4	0
2.4	20
2.0	50
1.8	75

Cooked cured meat products

Part of the initial (input) nitrite content is lost when the product is cooked (34–72% loss, depending on the heating conditions). Further loss occurs

Table 3.2 Initial composition and shelf-life of vacuum-packed bacon (data from Ranken, 1984)

Initial composition			Storage temperature			
			15°C (59°F)		25°C (77°F)	
Nitrate (ppm $NaNO_3$)	Nitrite (ppm $NaNO_2$)	Salt-on-water (%)	Days for nitrite content to fall to 10 ppm	Days to off-odour	Days for nitrite content to fall to 10 ppm	Days to off-odour
0	100	2.0	14	14	7	10
500	100	2.0	23	23	10	7
0	200	2.0	14	28	7	7
500	200	2.0	14	35	3	7
0	50	3.5	7	14	7	7
500	50	3.5	14	14	7	7
0	100	3.5	14	23	7	7
500	100	3.5	23	44	7	7
0	200	3.5	14	28	7	7
500	200	3.5	23	23	10	14
0	50	5.0	14	28	7	10
500	50	5.0	23	44	10	16

slowly and to a variable extent on storage; it is not possible to quote reliable figures for this.

There is an additional preservative effect when nitrite and protein are heated together: the 'Perigo effect', named after J. Perigo who discovered it. This effect is small and has not yet been clearly expressed numerically.

Special problems of nitrite

When considering problems of nitrite it is sometimes necessary to take account also of nitrate in so far as this may provide a source of nitrite.

Toxicity

Nitrite in moderate doses is toxic. It reacts with blood to form nitrosyl haemoglobin in the same way as it forms nitrosyl myoglobin in meat. The lethal dose is about 1 g for an adult. Its use is therefore restricted. In the UK only the following are permitted:

- sodium or potassium nitrate (E Nos E251, E252)
- sodium or potassium nitrite (E249, E250)

The permitted levels for nitrate and nitrite in cured meats in the UK are shown in Table 3.3.

Table 3.3 Permitted levels of nitrate and nitrite in cured meats in the UK

	Total nitrite + nitrate (mg/kg)*	Maximum nitrite (mg/kg)*
Cured meat in sterile packs	150	50
Acidified or fermented cured meats	400	50
Bacon and ham, uncooked or cooked but not sterile packed	500	200
Other cured meats	250	150

* Expressed as $NaNO_2$ in the final product.

Nitrited salt

In most countries now, nitrites for food use may only be supplied in admixture with salt. The quantities required for curing are therefore obtained by adding the appropriate amounts of nitrited salt. Styles in current use are:

Prague Salt	0.6% sodium nitrite in sodium chloride
Pökelsalz (Germany)	0.6% (was 0.4% for a time)
Sel nitrité (France)	0.6%
Nitrited salt (UK)	Various concentrations, as specified by the user; 50% in many cases

Nitrosamines

Nitrites react with certain amines (e.g. amino acids present in meat) to form traces of nitrosamines, the majority of which have been shown to cause cancers in animals.

Practical objectives, to be pursued as far as possible, are therefore:

- to ensure that meat products contain zero or minimum quantities of pre-formed nitrosamines and yield zero or minimum quantities on cooking;
- to restrict the amount of residual nitrite in meat products to the minimum technologically necessary, thus restricting the quantity available to form nitrosamines either on cooking or after ingestion by the consumer.

Nitrosamines pre-formed in the product or on cooking

Most meat products have been shown to contain or to form little or no nitrosamine. Exceptions are cured products containing fat, cooked to relatively high temperatures, especially in the presence of air. The most important cases are

- frankfurters and similar sausages (small amounts of nitrosamines produced)
- fried bacon, especially if fried to the crisp condition (moderate to large amounts of nitrosamines produced).

For bacon it has been shown that nitrosamine formation depends upon:

- input nitrite concentration (which probably affects formation of some precursor)
- residual nitrite concentration at the time of frying
- frying time × temperature
- fat content (small effect).

Nitrosamine formation may be reduced by the use of ascorbate or other antioxidants – but note the adverse effects of ascorbates in unpasteurised bacon, page 73.

Restriction of residual nitrite content

The residual nitrite content is affected by

- input nitrite content
- manufacturing process, especially any heating

- storage time and temperature
- pH of product
- presence of ascorbate, sulphur dioxide, etc.

Note also that there may be considerable variation between different batches of meat.

The relationships among the above conditions are complex, and fine control is difficult.

Nitrite and the formation of cured colour

This is discussed in Chapter 4.

Nitrate

The commonly used forms of nitrate are Chile saltpetre (sodium nitrate) and Bengal saltpetre (potassium nitrate).

The curing effects were once thought to be due to the saltpetre; it is now known that they are in fact due to nitrite: in curing brines, nitrate serves as a source of nitrite.

$$NO_3^- \xrightarrow{\text{reduction by bacteria}} NO_2^-$$

This change requires the presence of the appropriate bacteria, mainly micrococci and lactobacilli. If they are not present, or if other types are present in greater numbers, the desired change may not take place (page 162). The change is faster at higher temperature; nitrate is therefore especially useful as a reserve source of nitrite in bacon which may be subjected to high temperatures during storage and distribution (e.g. some van sales operations in summer weather).

Nitrate itself appears to have a small preservative action but this can probably be neglected in practice.

Small amounts of nitrate may be found in meat which has not been cured. Nitrate contents of some samples of untreated pork meat are shown in Table 3.4. Figures for nitrate found in uncured tongues are given on page 165.

Table 3.4 Nitrate contents of uncured pork meat (data from Ranken, 1984)

Nitrate content as ppm KNO_3	No. of samples
0	13
1–30	23
31–60	7
61–80	2
81–90	1

Ascorbate and erythorbate

(Erythorbic acid and the erythorbates are optical isomers of ascorbic acid and the ascorbates, with exactly the same chemical properties except that erythorbic acid and the erythorbates have no vitamin C activity. Erythorbates and ascorbates may therefore be used interchangeably. The use of erythorbates in meat technology was at one time prohibited in the UK but they are now permitted.)

These substances are used for enhancement or accelerated production of cured colour, especially in cooked cured products. They are of little value in uncooked products and may be detrimental in some circumstances (page 73).

Note that their use will reduce residual nitrite contents; since they have no antimicrobial action themselves, this will reduce the overall antimicrobial effect of the nitrite.

Phosphates

Preservative action

Some polyphosphates have a small preservative action additional to that of sodium chloride, especially when heated with nitrite, e.g. in cooked cured meats. The latter effect is not significant at low nitrite content levels, so phosphate cannot be used to replace nitrite.

Water binding, etc.

Phosphates which catalyse the salt effect are effective at about 0.3%. Some consider the taste unpleasant at this level.

Heat

Heat may be applied in various ways, apart from cooking just before consumption. Products may be:

- pasteurised in cans, vacuum packed or in other containers
- pasteurised but not in sealed containers
- sterilised in cans or other sealed containers
- hot smoked.

Heating to temperatures of 55°C (131°F) or more (cured meats) or 65°C (149°F) or more (uncured meats) has a preservative action by destroying or inactivating most spoilage and harmful organisms (though not *Clostridium botulinum* spores). The size of the effect depends on other factors (salt-on-water content, nitrite content, phosphates present, etc.) in a complex manner.

See page 113 for calculations in the case of pasteurised cured meats. Similar principles may be applied to other cases.

Smoke

As well as its strong flavour, smoke has a preservative effect and where the smoking process also includes drying there is a further benefit of reduced water activity. See page 151 for more detail.

Storage temperature

The lower the temperature, the slower the microbial changes and the longer the shelf-life.

Vacuum-packed sliced bacon of average salt and nitrite content may be expected to keep satisfactorily for:

12 weeks at 0°C, 32°F
5–6 weeks at 5°C, 41°F
2–3 weeks at 15°C, 59°F
4–6 days at 25°C, 77°F

Unsliced sides or blocks of bacon, reasonably protected at the surface, should keep a little longer than these times.

4 Colour and Flavour

THEORY

If a narrow beam of white light is passed through a glass prism, it is split into a rainbow-like band of colour of different wavelengths showing that white light is in fact a combination of all colours of the spectrum.

An object appears coloured when some wavelengths of light are selectively absorbed. Fresh meat looks red because the natural colouring matters absorb all colours other than red, which is reflected.

When an object is viewed in reflected light its colour is dependent on:

- The nature of the illuminating light. Different illuminants contain different amounts of light of different wavelengths. If meat is viewed in an illuminant which has a proportion of red wavelengths, e.g. tungsten light, it will appear more red because more red light is available to be reflected to the eye.
- Changes taking place during reflection. These changes are related to:
  - the nature of the pigments present in the meat (factors affecting the chemical and physical structure of meat pigments are discussed later);
  - the amount of light reflected from below the surface of the meat, which is partly dependent upon the physical structure of the meat; for example, if soluble protein has been precipitated, light scattering will occur and the meat will therefore appear pale.

Iridescence

A multi-coloured iridescence is sometimes seen on cut surfaces of meat when viewed at certain angles. These colours are interference patterns caused by interaction of light waves with the regular fibrous structure of the meat.

MEAT PIGMENTS

Chemistry

The main colouring matter of meat is myoglobin. Depending on how well bled the animal has been, meat also contains a small proportion of haemoglobin – the pigment in blood. For practical purposes, myoglobin and haemoglobin are very similar; therefore only myoglobin is considered here.

Structure of myoglobin

Myoglobin consists of a protein (the globin) and a non-protein portion (the haem group) (Fig. 4.1).

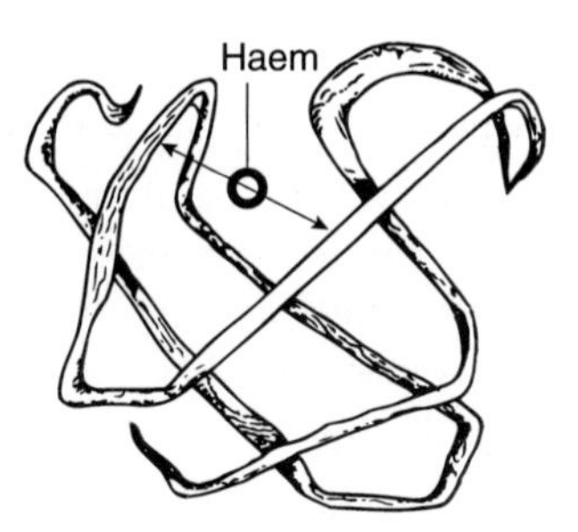

Fig. 4.1 The myoglobin molecule.

The haem group (Fig. 4.2) consists of a flat porphyrin ring with a central iron atom. The iron atom has six bonding points or coordination links. Four of these are linked to nitrogen atoms; one is attached to the globin molecule; the remaining linkage is free to bind to other substances, usually water or oxygen.

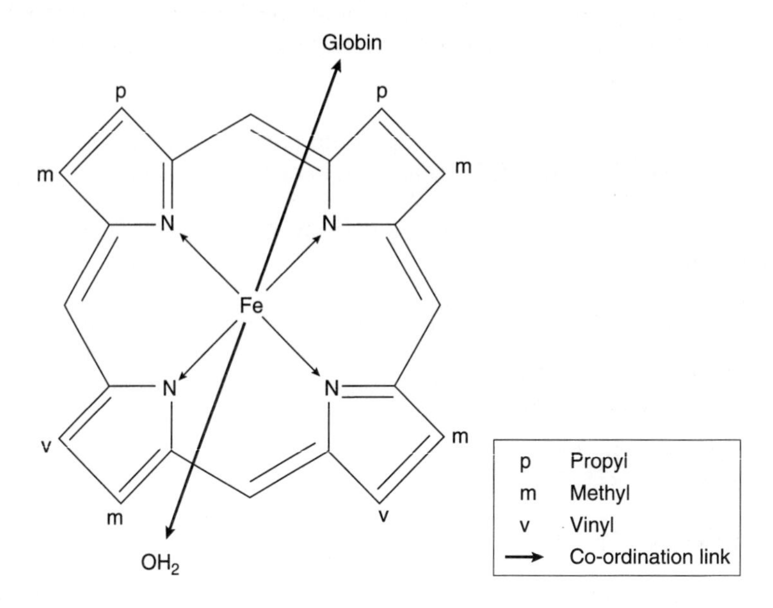

Fig. 4.2 The haem group in myoglobin.

Colour of myoglobin

The colour of the pigment depends on at least three factors:

- The oxidation state of the iron atom. It may be reduced, Fe^{2+}, or oxidised, Fe^{3+}
- The nature of the group at the sixth bonding point of iron.
- The state of the globin. It may be native as in raw meat, or denatured as in cooked meat. See Table 4.1.

Table 4.1 Haem pigments in uncured meat

Pigment	Oxidation state	Sixth linkage	State of globin	Colour
Reduced myoglobin	Fe^{2+}	H_2O	native	purple
Oxymyoglobin	Fe^{2+}	O_2	native	bright red
Metmyoglobin	Fe^{3+}	H_2O	native	brown
Denatured globin haemichrome	Fe^{3+}	H_2O	denatured	brown

In raw meat and meat products the desired colour is usually the bright red oxymyoglobin. In cooked meat and meat products the desired colour is usually the brown denatured globin haemichrome.

In extreme conditions, the pigment may be decomposed; the haem portion becomes detached from the protein; the porphyrin ring is disrupted and finally the iron atom is lost from the haem structure. Green choleglobin and colourless bile pigments are formed.

COLOURS OF MEAT

There are four quite different conditions which must be clearly distinguished:

- **fresh** or **raw**, **uncooked** meat
- **cooked** meat
- **cured**, **uncooked** meat
- **cooked** and **cured** meat.

We shall consider these in turn.

COLOUR OF FRESH MEAT

Reduced myoglobin, oxymyoglobin and metmyoglobin are all present in fresh meat, in equilibrium with one another.

In the centre of a piece of meat there is no oxygen and the pigment is in the purple reduced myoglobin form (the same colour appears in vacuum-packed meat, e.g. primal cuts). At the surface of a piece of meat there is a good oxygen supply and bright red oxymyoglobin is formed. Between these two zones is a region of low oxygen concentration, which favours oxidation of pigment to metmyoglobin. A layer of brown metmyoglobin therefore forms just below the surface of the meat. These layers of colour are shown in Fig. 4.3.

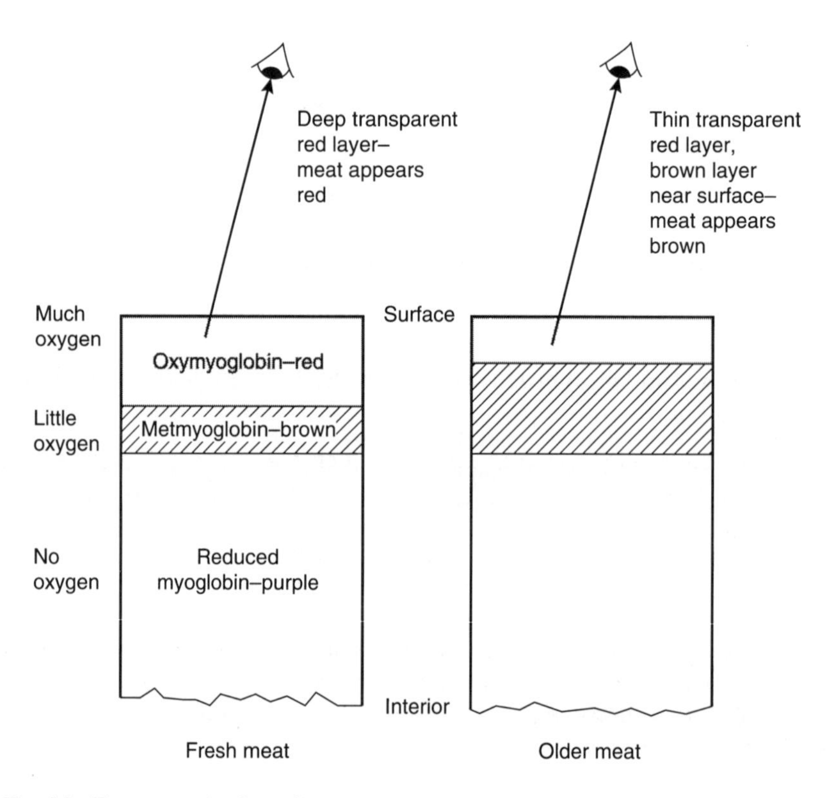

Fig. 4.3 The apparent colour of meat.

Chemistry

The relative proportions of the three pigments depend on conditions within the meat. Metmyoglobin cannot take up oxygen but enzymes present in the fresh meat are capable of reducing metmyoglobin to reduced myoglobin, which can then take up oxygen to form red oxymyoglobin. As the meat ages, the substrate for these enzymes is gradually used up; metmyoglobin can no longer be reduced and the brown metmyoglobin layer becomes wider until it finally becomes visible through a narrowing oxymyoglobin layer and the meat appears brown. (See Figs 4.3 and 4.4.) This is a simplified picture; other factors such as bacterial load (page 64) and dehydration play a part.

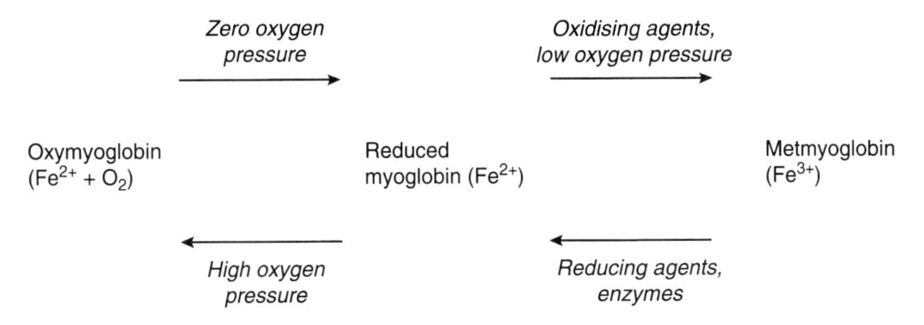

Fig. 4.4 Chemical relationships among fresh meat pigments.

Factors affecting the colour of fresh meat

Pigment concentration

This varies greatly among different meats. The following factors are important.

- Species: beef, for example, contains much more myoglobin than pork (Table 4.2).
- Breed.
- Age: pigment concentration increases with age.
- Sex: meat from male animals usually contains more pigment than that from female animals.
- Muscle function: the function of myoglobin is to store oxygen; therefore, muscles which do more work contain more myoglobin, e.g. leg muscles are deeper red than loin.
- Variation within muscles.

Table 4.2 Myoglobin concentrations in meat

Meat	Myoglobin concentration (mg/g wet tissue)
Beef	4–10
Veal	*c.* 3
Lamb	3–7
Pork	2–7
Poultry	
dark meat	2–3
white meat	0–0.5

Oxygen concentration and packaging

Oxygen-permeable film

Maintenance of the bright red oxymyoglobin colour depends on an adequate supply of oxygen. Packing films must therefore have high oxygen permeability. Five litres of oxygen per m^2 per day are required. Coated cellulose films and low-density polythenes have high oxygen permeability but low water permeability and are satisfactory.

Vacuum packs

No oxygen is present; therefore myoglobin is in the purple, reduced form. The meat will retain this colour for long periods in vacuum pack and this method is used for primal meats. Bacterial growth is restricted under vacuum conditions, which is another advantage. The purple colour is not generally acceptable to retail consumers but the bright red colour of oxymyoglobin is restored soon after the pack is opened; the meat can then be re-packed for retail sale.

Oxygen-containing packs

The meat is packed in a gas-impermeable pack containing a high concentration of oxygen. Colour quality can be maintained for up to 14 days in the right conditions. Carbon dioxide gas may be added to suppress bacterial growth; various combinations of gases are possible. There may be a double packaging, with an outer layer to provide the necessary impermeability to gases. This is removed before retail display, leaving an inner, oxygen-permeable layer.

Microbiological effects

Reduction of oxygen concentration

Aerobic bacteria consume oxygen, thus reducing the oxygen concentration and causing browning. This is particularly important in minced meat, which has a large surface area.

Products of bacterial metabolism

- Hydrogen sulphide products. Some bacteria produce hydrogen sulphide (H_2S); this combines with myoglobin to form green sulphmyoglobin. This is the cause of 'greening' in uneviscerated poultry and in some over-aged vacuum-packed meats.

- Hydrogen peroxide products. Hydrogen peroxide (H_2O_2) is a strong oxidising agent which causes pigment breakdown and greening or pale colours.
- Coloured by-products. Some *Pseudomonas* species cause blue/green discoloration.

Coloured bacteria

Some *Sarcina* or *Micrococcus* species cause red discoloration.

Temperature

Chill conditions

The rate of oxidation of the pigment to metmyoglobin increases with increasing temperature. Red colour is therefore more stable at lower temperatures. At low temperatures, the solubility of oxygen is greater and oxygen-consuming reactions are slowed down. Therefore, there is a greater penetration of oxygen into the meat and the meat is redder than at high temperatures.

Frozen conditions

Rapid freezing results in the formation of small ice crystals which cause a lot of light scattering, giving the meat a pale, opaque appearance. Slow-frozen meat contains large ice crystals, which scatter less light, so that the meat has a dark, translucent appearance. These colour changes disappear on thawing out. Meat has a better frozen colour if allowed to 'bloom' in air before freezing.

Storage conditions

Metmyoglobin formation is maximal at about −12°C (+10°F). This is probably because at this temperature only part of the water is frozen; salts are concentrated in the unfrozen portion and the high salt concentration promotes pigment oxidation. Discoloration is greatly accelerated by light. The red colour is best preserved by storing in the dark at −18°C (0°F) or below.

Freezer burn Ice sublimes from unprotected areas of the meat, leading to desiccation, denaturation of proteins and oxidation of the pigment. To prevent this, meat should be lightly wrapped in undamaged moisture-impermeable film.

Thawed meat

The colour of thawed frozen meat is less stable than that of fresh meat. Meat which has discoloured during frozen storage will remain brown after thawing.

pH

DFD (dark, firm, dry) meat

This meat has a high ultimate pH. At this high pH the muscle fibres are swollen with water and tightly packed together. Oxygen penetration is low. Also, surviving respiratory enzyme activity is high. As a result the oxymyoglobin layer is narrow and the purple myoglobin layer shows through. There is little light scattering at the surface of the meat and the meat appears dark (dark-cutting beef; glazy bacon). (See page 10).

PSE (pale, soft, exudative) meat

The pH falls while the carcass is still warm, causing partial denaturation of the proteins and an increase in the amount of light scattered; part of the pigment is oxidised and the meat appears pale. (See page 10.)

Light

Light has little direct effect on the colour of **fresh meat** at chill temperatures but care should be taken that high illumination does not cause a rise in temperature due to a 'greenhouse' effect in the package. Even in a chill cabinet, light energy absorbed through the transparent film may cause the product temperature to rise above that of the ambient air and the meat may therefore become warm.

Ultra-violet light causes protein denaturation, which will result in browning in the longer term.

In **frozen meat**, light accelerates discoloration. Products should therefore be covered during frozen storage, as a precaution against the normal lighting in the cold store.

Reducing agents

Reducing agents convert metmyoglobin to reduced myoglobin. This form of the pigment readily takes up oxygen to form bright red oxymyoglobin. Reducing agents therefore increase the colour stability of fresh meat.

In principle, chemical agents are not permitted in fresh butcher's meat. Because colour is an indicator of freshness, meat with added reductants and therefore with improved colour might look good when it is in fact bacterially unsound. However, there are approved reducing agents which may be used in meat products in certain circumstances:

Ascorbic acid and its sodium salt

These are the most common reducing agents; a level of 200–500 ppm is effective. Erythorbic acid and the erythorbates are optical isomers of ascorbic acid and the ascorbates, have the same chemical properties and may be used in their stead; they are usually cheaper.

They are effective in the absence of air. In the presence of air, high concentrations of ascorbate (over 1000 ppm) cause the formation of hydrogen peroxide which destroys the haem structure and results in green or bleached products.

Nicotinic acid and nicotinamide

Nicotinic acid and nicotinamide form red compounds with reduced myoglobin and are therefore most effective when used in combination with ascorbate in the absence of air. In vacuum-packed meat, 600 ppm of nicotinic acid is effective.

Nicotinamide, though not nicotinic acid, also promotes the formation of metmyoglobin, therefore enhancing colour formation, by preventing the destruction of nicotinamide adenine dinucleotide (NAD) by nucleosidase. In the presence of air, however, both substances increase the rate of browning.

Sulphur dioxide

Sulphur dioxide is a permitted preservative in the UK. It is also a reducing agent and improves the colour stability of sausages, burgers, etc., at the level permitted (450 ppm).

Oxidising agents

Oxidising agents promote the formation of brown metmyoglobin and are therefore to be avoided. Note particularly the following:

- Cleaning agents and bleaches often contain oxidising materials and should not come into direct contact with meat.
- Ozone from arc welding equipment has been known to cause discoloration.
- Rancid fats contain peroxides which are strong oxidising agents. Note that frozen fat may increase the problems (page 94).

- Nitrite is an oxidising agent. A level as low as a few parts per million is sufficient to cause discoloration. Great care should be taken to avoid cross-contamination between fresh and cured meat operations.
- Any factor which causes denaturation of the globin part of the pigment (e.g. high salt concentrations, dehydration) will also tend to oxidise the haem part.

Metals

Metal ions, especially copper, promote autoxidation of oxymyoglobin to metmyoglobin; iron and zinc have less effect. Sequestering agents (e.g. citrate or phosphate) improve colour stability.

COLOUR OF COOKED (UNCURED) MEAT

Chemistry

The pigment responsible for the brown or grey colour of cooked meat is denatured globin haemichrome. The iron atom is in the ferric state, Fe^{3+}, and the globin is denatured.

Denaturation of myoglobin in meat occurs at temperatures from about 60°C (140°F) upwards. The process usually appears to be complete (disappearance of red colour) in poultry meat at 67°C (152°F), in pork at about 70°C (158°F) and in beef at 75°C (167°F) or even sometimes as high as 85°C (185°F).

Premature browning

This unfortunate phenomenon is sometimes observed when ground meat products such as hamburgers are cooked. Contrary to the information above, the meat may show a brown colour when cooked to temperatures as low as 55°C (131°F). This can occur when fresh meat has been used for the manufacture and the product is cooked within only a few hours when the meat is still in an oxygenated or oxidised state, with a high proportion of the pigment present as oxymyoglobin. Obviously in such cases the heat treatment can be insufficient to destroy pathogenic micro-organisms and ensure safety. With sufficiently aged meat where the majority of the pigment is in the metmyoglobin form, the problem does not occur.

Red or pink colours in cooked uncured meat products

The most likely cause of red colours in cooked meat, especially near the centre of the meat portion, is of course undercooking – failure to reach the denaturation temperature of the myoglobin. So a residual red colour is usually taken as an indication that the meat has not been cooked enough for microbiological safety.

However there are two important exceptions to this rule:

- In beef, where the denaturation temperature, as noted above, is 75°C (167°F) or higher, the red appearance of 'rare' cooked meat will not necessarily mean that the meat is undercooked in respect of its microbiological safety.
- After meat has been fully cooked to a brown colour, a red colour may be found later on cooling; this red colour reappears at the centre of the meat or meat product where conditions are anaerobic and is due to the conversion of the brown ferr*i*haemochrome pigment to the red ferr*o*-haemochrome.

Other possible reasons for red or pink discoloration are:

- Nitrite contamination. Small amounts of nitrite may result in the formation of pink nitrosyl myoglobin during cooking. Again, cross-contamination between fresh and cured meat should be avoided.
- Nitric oxide, e.g. from combustion of natural gas or coal gas, has a similar effect.
- Nitrate is often present in mains water. Although nitrate cannot directly discolour meat, it may sometimes be reduced to nitrite. The problem arises when meat remains in contact with the nitrate, before heat treatment, for sufficient time for reduction to take place (e.g. if the appropriate bacterial flora or chemical conditions are present). Under appropriate conditions, less than 10 ppm nitrate can be converted to sufficient nitrite to cause pink discoloration.
- Carbon monoxide forms a stable deep red pigment with myoglobin: carboxymyoglobin.
- The use of polyphosphates is sometimes associated with pink or yellowish colours. The causes are not fully understood but might in some cases be associated with nitrate or nitrite impurity in the phosphate.

COLOUR OF CURED MEAT, UNCOOKED AND COOKED

Chemistry

Colour formation and the changes which take place are shown in Fig. 4.5.

Nitrosylmyoglobin is responsible for the red colour of raw cured products. It is formed from the reaction of myoglobin with nitric oxide.

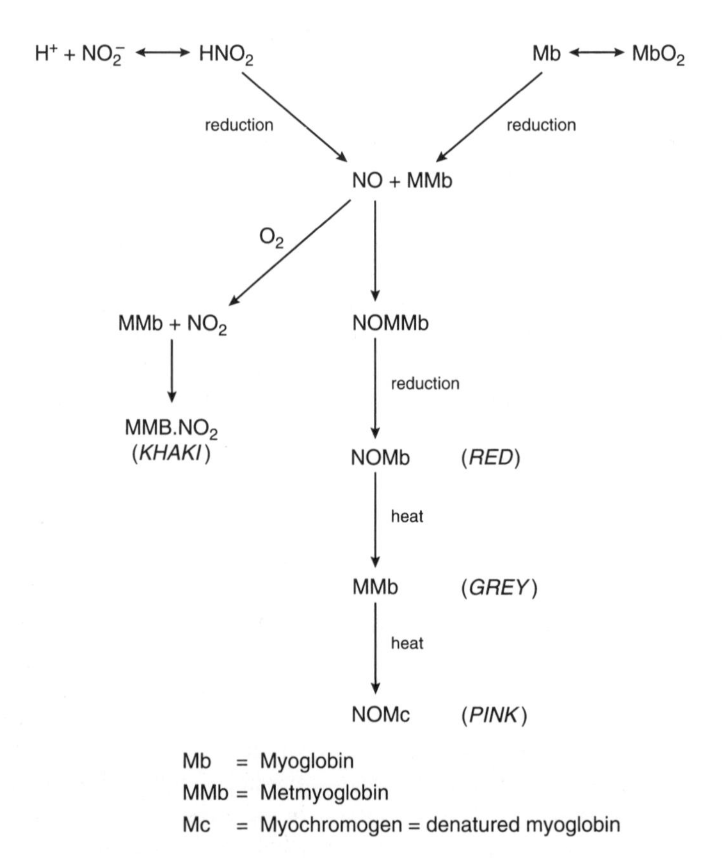

Fig. 4.5 Colour changes in cured meat.

Sodium nitrite is the usual source of the nitric oxide – in solution the nitrite ion exists in equilibrium with undissociated nitrous acid:

$$HNO_2 \longleftrightarrow H^+ + NO_2^-$$

In acid conditions this equilibrium is pushed towards the left, which is beneficial to the reactions which follow. Nitrous acid is thought to decompose in slightly acid conditions in the following way to yield nitric oxide:

$$3HNO_2 \longrightarrow HNO_3 + H_2O + 2NO$$

Nitrite is also an oxidising agent which rapidly converts myoglobin (red) to metmyoglobin (brown). Nitric oxide then combines with metmyoglobin to form nitrosyl metmyoglobin, which is then reduced to nitrosyl myoglobin:

$$\text{Mb} \xrightarrow{+\text{NO}_2^-} \text{MMb} \xrightarrow{+\text{NO}} \text{NOMMb} \xrightarrow{\text{reduction}} \text{NOMb}$$

Conversion of myoglobin to the nitrosyl form is incomplete, and not very consistent; it may vary between about 35% and 75%, with input nitrite 100–150 ppm, in different samples of meat.

During heating, nitrosyl myoglobin is denatured to pink nitrosyl myochromogen. The exact structure of the molecule is not known. The iron atom remains in the reduced form and the globin portion is denatured. The change to the cooked pigment is sometimes preceded by a grey colour, which suggests that the nitric oxide molecule temporarily dissociates from the myoglobin during heating, resulting in an intermediate grey metmyoglobin stage. This has not, however, been confirmed experimentally.

During cooking more nitrosyl pigment is formed. Again the proportion converted is irregular, between about 60% and 90% on cooking at 70°C (158°F).

Residual nitrite is required at the time of cooking. Bacon with low or zero nitrite at the time of frying may be grey after cooking.

Factors affecting the colour of cured meat

Nitrite

The amount of nitrite required for colour development and stability in various experimental conditions is summarised in Table 4.3.

In practical conditions, the variability in cure distribution must also be taken into account. It is relatively easy to achieve even distribution of nitrite in a highly comminuted system such as luncheon meat, but in large pieces of meat such as bacon this is very difficult.

High concentration of nitrite, especially at low pH, can result in the formation of green nitrihaemin or 'nitrite burn'.

Nitrate

In traditional processes nitrate was the only source of nitrite:

$$\text{Nitrate} \xrightarrow[\text{(micrococci)}]{\text{bacteria}} \text{Nitrite}$$

In classical Wiltshire curing, the correct bacterial flora are maintained in the brine for many years by careful control of the conditions (page 155). In modern rapid-cure processes nitrite is added directly to the brine but a proportion of nitrate may also be included. Whether this nitrate is converted to nitrite depends on whether the appropriate bacteria exist in the factory environment, for example in a factory with a previous history of

Table 4.3 Nitrite requirements for colour formation and stability in bacon (data from Ranken, 1984)

Product	Colour formation	Colour stability		
	Input nitrite required (ppm)	Conditions (all at 5°C, 41°F)	Nitrite required ppm (input)	(at time of cooking)
Unheated				
Bacon	10–20 gives non-uniform colour; 50 satisfactory	2 weeks in vacuum pack, in light or dark	40–70	
		3 days out of pack in light or dark	70–100	
Heated				
Bacon	30 at time of cooking	3 weeks in vacuum pack, in the dark	20–30	10–15
		3 weeks in vacuum pack, in the light	30–40	15–20
Luncheon meat	10	Indefinite	10	

making Wiltshire bacon. If the appropriate bacteria are not abundant, as in a new factory, there may be no such conversion and the added nitrate will be ineffective or possibly detrimental (page 162).

Note that small concentrations of nitrate may occur in the uncured meat (page 56).

Reducing agents

Colour formation – raw products

Reducing conditions are necessary for the formation of nitric oxide and for the reduction of nitrosyl metmyoglobin. Enzymes and other reducing systems are naturally present in meat but they can be supplemented by adding chemical reducing agents. Ascorbate is the reducing agent most commonly used.

Ascorbate is believed to improve the efficiency of curing by taking part in a reaction with nitrous acid:

$$2HNO_2 + \underset{\text{(ascorbic acid)}}{C_6H_8O_6} \longrightarrow 2NO + 2H_2O + \underset{\text{(dehydroascorbic acid)}}{C_6H_6O_6}$$

This reaction represents a saving of approximately one-third of the nitrite. Ascorbate also helps to remove traces of oxygen, which inhibits cured colour development.

In practice the effect of ascorbate varies with different batches of meat. This is probably partly due to variation in the natural reducing ability of meat. In general, ascorbate increases the rate of colour development and the amount of cured pigment formed; it improves the intensity and uniformity of colour at low nitrite levels.

Colour formation – cooked products

Ascorbate slightly increases the amount of nitrosyl pigment formed during cooking. The difference is not usually big enough to affect the colour intensity when adequate nitrite is present (100 ppm). Residual nitrite is necessary at the time of cooking for a good cooked colour (Table 4.3). Since ascorbate speeds up the breakdown of nitrite, the residual nitrite in the product decreases more quickly in the presence of ascorbate. In bacon, which may be stored for relatively long periods before cooking, ascorbate may be detrimental because the nitrite level may be reduced below the minimum required for a good cooked colour.

Colour stability – raw products

Ascorbate improves the stability of raw vacuum-packed bacon. In the presence of air, however, ascorbate can cause rapid loss of colour.

Colour stability – cooked products

Ascorbate is generally considered to improve the colour stability of cooked cured products stored in air and vacuum pack. Ascorbate improves the colour stability in air of cooked pork mixtures containing adequate nitrite (at least 20 ppm). On the other hand, ascorbate does not affect the rate of colour loss but there is an improvement in initial colour intensity due to ascorbate and this is maintained throughout storage.

The main effects of ascorbate (or erythorbate) are summarised in Table 4.4. Note that variable results may be obtained if the nitrite and ascorbate are unevenly distributed in the product.

Practical note

Sodium ascorbate (or sodium erythorbate) should always be used in curing brines. **Do not use ascorbic *acid* or erythorbic *acid*.**

At the low pH of ascorbic acid, nitric oxide is formed so quickly that it may be released as a gas. Nitric oxide combines with oxygen in the air to

Table 4.4 Effects of ascorbate in cured meats

	Colour formation	Colour stability
Raw products	(i) Rate of formation increased	(i) Improved in vacuum pack
	(ii) Intensity and uniformity improved, especially at low* nitrite levels	(ii) Colour destroyed by excess ascorbate in the presence of air
Cooked products	Improved at low* nitrite levels	Improved in air and vacuum pack

* Usually below 30 ppm input. See also Table 4.3.

form thick brown fumes of nitrogen dioxide. Nitrogen dioxide is highly toxic; it also causes a khaki-brown discoloration of the meat – see Fig. 4.5.

Air

The presence of air inhibits the formation of both raw and cooked cured colour. Both raw and cooked colours are unstable in air. The shelf-life of cured products is greatly increased by vacuum packaging.

Light

Light catalyses the dissociation of nitric oxide from cured meat pigment. Both raw and cooked cured products rapidly become brown when illuminated, especially in the presence of air. Exposure of cured products to light should be avoided whenever possible. For display purposes, illumination levels should be kept low and careful rotation of stock should be practised.

Temperature

Cooking temperature

The transition from the raw to the cooked colour is complete at about 70°C (158°F). Nitrosyl pigment formation is also maximal at about this temperature.

Storage temperature

Above the freezing point the rate of discoloration of cured products increases with increasing temperature. For special problems in frozen meat see page 94.

pH

The rate of development of cured colour is faster at low pH. The stability of the colour once formed, however, is greater at higher pH.

Oxidising agents

Oxidising agents tend to promote discoloration of cured meats. Hydrogen peroxide formed by bacterial metabolism is particularly important in cured meats. This is because the enzyme catalase, which normally destroys peroxide in fresh meat, is deactivated by the high salt concentration of cured meats. Fat peroxides also promote discoloration (page 94).

Metals

Iron forms black iron sulphide, which is a common cause of black spots or patches in many foods.

Copper has been implicated in black discoloration of cured meats. For instance, 5 ppm copper in a curing brine has caused intense black discoloration in canned tongues.

ADDED COLOURS

Colouring matters are sometimes used to supplement the natural colour of fresh and cured meat products. They fall into four categories: artificial, natural, nature-identical and inorganic.

Red colours of possible use in meat products and permitted in the UK are shown in Table 4.5.

Table 4.5 Permitted colours, suitable for meat products (UK Regulations, 1995) (Some colours are permitted only in certain products)

Colour	EU No.	Type
Curcumin	E100	Natural
Carmine (Cochineal)	E120	Natural
Ponceau 4R	E124	Artificial
Red 2G (Geranine)	E128	Artificial
Allura Red	E129	Natural/artificial
Caramel (all types)	E150 a–d	Nature-identical
Carokenes	E160 a	Natural
Capsanthin, Paprika extract	E160 c	Natural
Beetroot red (Betanin)	E162	Natural

Simulation of uncured meat colour

Red 2G is the most common artificial colour added to British sausages. Unlike most of the other permitted red colours it is stable to sulphur dioxide and exposure to light during storage.

Beetroot red at 1000 ppm gives an acceptable colour in the absence of sulphur dioxide but the colour is destroyed if sulphur dioxide is present.

Roxanthin (a modified Canthaxanthin; not currently a permitted additive in the UK) gave a satisfactory colour which was stable to sulphur dioxide.

Simulation of cured meat colour

Ponceau 4R may be used to imitate the colour of uncooked bacon, but it is difficult to colour slices of meat uniformly.

Attempts have been made experimentally to produce products resembling bacon or luncheon meat without nitrite but with preservative added colouring matter instead. Carmoisine and Ponceau 4R were tried in bacon, Red iron oxide, Amaranth, Carmine, Carmoisine, Erythrosine, Ponceau 4R and Red 2G in luncheon meat. None was found satisfactory and, in any case, none is now permitted in such products in the EU.

MISCELLANEOUS COLOUR PROBLEMS

Two-toning

This is a condition sometimes seen in hams where adjacent muscles appear as different shades of pink. This is a result of natural variation in pigment levels between the muscles; it can be remedied only by selecting similar-looking muscles beforehand.

Onions

Onions may be implicated in the browning of meat products. This is usually due to pyruvic acid, the component responsible for pungency of the onion which is also a strong oxidising agent. Residual enzyme activity may also play a part. This problem can occur with fresh, frozen or dried onions. It may be reduced if the onions are pre-cooked, or by selecting less pungent varieties.

Sausages

Various colour defects in British fresh sausages are referred to on page 141.

Alternatives to nitrite for formation of the cured colour

Various possibilities to eliminate the use of nitrites altogether and thus reduce possible hazards from nitrosamines were investigated in the 1970s. Alternative preservatives such as sulphur dioxide, nisin and sorbic acid were tried, along with natural and artificial food colours. Attempts were also made to produce nitrite-free bacon by using nitric oxide gas – British Patent No. 1375700. None of these alternatives was more than moderately successful under laboratory conditions, and none has been developed commercially.

FLAVOUR

The flavour of a meat product derives from four main sources.

The flavour of the meat itself

There has been a great deal of research over the years into the chemical components of meat flavour, and a very large number of substances have been found to be involved. It is not essential for the practising technologist to know all the details, for there is little that can be done to alter the inherent flavour of meat after the animal has been slaughtered. Note however, the general points:

- 'Meat' flavour increases with the age of the animal at slaughter, so hen meat has more flavour than young chicken, and mutton has more than lamb.
- The characteristic flavours of beef, lamb, chicken, etc. reside more in the fat than in the lean meats; 'fat-free' or fat-reduced products are likely to be less flavoursome than those with higher fat content.

The flavour of a meat or meat product may be changed by other factors listed below.

Spoilage

Flavour may be more or less affected by microbiological or by oxidative chemical changes. See pages 53, 88 for microbiological changes, page 94 for oxidative changes.

Flavours developed during processing

The characteristic flavours developed in the course of curing or fermentation processes are referred to under the appropriate headings.

Added flavours

Most of the curing and processing agents used in meat technology, such as salt, phosphates or smoke, have characteristic flavours, the effects of which are dealt with under the appropriate headings.

Herbs, spices and flavouring substances, including spice extracts, smoke extracts, flavour enhancers (such as MSG) and some recipe ingredients (such as onion), all have uses in particular meat products. The quantities used are usually small, of the order of 0.1%. The recipes for their use may be traditional to particular kinds of product, or may be considered unique and secret to particular manufacturers; it is not proposed to comment further here on this aspect.

5 Microbiology

Because microbial growth in meat products can cause so many difficulties, it is essential to be aware of the main possibilities and to know what to do about them in the factory.

However, this is not a microbiological textbook. Details of microbial types, methods of detection and measurement and much else are not included. In many practical cases, these details may be highly important and a specialist would be needed to deal with them properly. The help of a skilled microbiologist should therefore be sought whenever necessary.

PRACTICAL: THE MAIN THINGS TO KNOW

(1) **Microbes are always present on fresh meat** and most other food ingredients. It must be borne in mind that any or all of the troublesome types may be present, even if initially they are in small numbers.

(2) **Microbes grow readily on meat.** As the numbers of microbes increase, the following effects appear in turn:

- brown or grey colour in the uncooked meat (especially burgers, fresh sausages) (page 64);
- spoilage smells;
- slime, mould growth, etc. (visible presence of the microbes).

(3) **Microbial growth is the main cause of spoilage.** The **storage life** of most meat or meat products is the time it takes to grow sufficient numbers of microbes to cause the discoloration, smell, etc. (There are other causes of spoilage but this is the most common.)

(4) **The more the meat is cut the greater the microbial growth.** There is some protection at first from skin or intact muscle sheaths; the more the meat is exposed by cutting, the more food is made available to the bacteria and the more surfaces there are for them to grow on. Thus, when all else is equal, uncut sides will always have lower counts than minced meat.

(5) **Temperature affects growth:**

- At ordinary working temperatures, **the higher the temperature the greater the growth.**

  - **The microbial danger zone** (the temperatures of greatest growth) is 10–63°C (50–145°F).
  - **Working temperatures** are controlled by law in most countries: in the EC and UK, meat products and meat intended for cutting should be held at not more than 12°C (54°F).

- Growth stops in frozen meat (but only below −10°C, +14°F, for mould) but only a few of the microbes are killed. The majority will survive and grow again when the meat thaws.
- Above 70°C (158°F), held for 2 minutes, meat is **pasteurised** (free from active microbes, but may still contain spores which can begin to grow again on cooling).
- Meat is **sterilised** (with the spores also killed) when heated to at least 100°C (212°F) **for several hours**, or to higher temperatures for shorter times.

(6) **Microbial growth is a major cause of food poisoning** (page 86).

MICROBIAL GROWTH

Effect of acidity (pH)

As a general rule, the microbes which cause meat spoilage or food poisoning grow better under low acid conditions (higher pH) than under higher acid conditions (lower pH). Exceptions include the lactobacilli which are significant in special cases, e.g. bacon brines (page 155), and fermented sausages (page 167).

In addition, preservatives such as sulphur dioxide or nitrite are more effective at higher acidity (lower pH).

Conditions which give low acidity (high pH) should therefore be avoided if possible, for example excessive use of alkaline (high pH) phosphates. Where such conditions cannot be avoided, it may sometimes be noted that shelf-life, for instance, is shortened. For example, collar bacon, with higher pH, usually has a shorter shelf-life than back bacon.

Moisture and mould growth

Moulds grow well in damp but not wet conditions, e.g. RH 85–95% (some down to RH 65%). As with other microbes, they are killed at pasteurising temperatures. Some species will grow under freezing conditions, growing slowly down to about −10°C (+14°F) (page 92).

Damp conditions commonly result when condensation occurs on an otherwise relatively dry surface. Condensation is therefore likely whenever a warm atmosphere comes into contact with a cool surface. If the surface consists of or contains food material for the moulds, mould growth can occur.

Mould growth inside pies

See page 181.

Condensation and 'structural mould' outside cold stores, etc.

Even when well insulated, the walls of a cold store are usually cooler than the surrounding atmosphere. If the walls are clad with absorbent materials, e.g. plywood, the condensation may not be visible but dampness may be retained and mould growth can occur. This is unsightly and a possible source of infection of products.

The problem may be cured by installing impermeable cladding, e.g. aluminium or hard plastic (mould-resistant paint may not be very effective).

Condensation is always likely on goods brought out of cold store into warmer atmospheres.

Unwanted microbial growth during manufacturing processes

This is especially likely in jelly solutions for pies (page 178), solutions of flavours, phosphates, etc., for injection into poultry or other meats, and can-cooling water, when any of these is recycled. Note that:

- some of the liquor may remain for long times in pumps, valves, pipes, etc.;
- the overflow liquid is likely to be contaminated with juice, etc., from the product being treated;
- the composition of the liquor, especially if contaminated by the product, may encourage microbial growth (note that curing brines with high salt content are normally fairly safe in this regard, but most other common liquors are not);
- temperatures may not be controlled.

Microbial growth is a strong possibility in these conditions. If it occurs, liquor used later in the day will contaminate the product made then.

Action to take:

- temperature control – wherever possible, liquor should be either cold (below 10°C, 50°F), or hot (above 75°C, 187°F);
- regular emptying of the system, with thorough cleaning.

Positive uses of microbial growth

Curing brines

See page 50 for the use in the Wiltshire bacon process of 'live' brines, i.e. brines in which active, controlled, microbial growth takes place.

When micrococci and certain other bacteria are present, plus soluble solids from meat immersed in the brine, plus salt concentration nearly saturated or saturated (26% w/w, 31% w/v):

$$\text{nitrate} \xrightarrow{\text{microbial growth}} \text{nitrite}$$

The **nitrite** content of the brine is thus maintained by adding more **nitrate**.

If the same microbes are present on the product (e.g. bacon), then nitrate is converted to nitrite in the product. That is, the nitrate content of the product provides a reserve supply of nitrite and therefore increases storage life. If the necessary microbes are not present on the product, nitrate will not be converted to nitrite.

Similar principles apply to brines used for other products.

Fermented sausages

Microbial changes in fermented sausages etc. are due to microbes naturally present in the manufacturing plant, or more usually in modern systems, to added starter cultures. The main microbial processes are:

(a) Nitrate $\xrightarrow{\text{micrococci}}$ nitrite

(b) Carbohydrate (sugars, etc.) $\xrightarrow{\text{lactobacilli}}$ lactic acid

(c) Surface mould growth may be allowed or encouraged during a final, long, drying-out process.

The final combination of lactic acid content, low moisture content (= low water activity) and residual nitrite content gives long shelf-life at ambient temperatures.

Management of these changes, to the right degree and in the right sequence, requires care and skill. Fuller details are given in Chapter 9.

Control or destruction of microbes

Preservatives

These are substances which slow down or prevent microbial growth, thereby prolonging shelf-life. In all countries their use is severely restricted. The principal ones permitted in meat products are listed below.

- **Nitrate and nitrites in cured meats.** See page 51 for discussion of the preservative effects and the problem of ensuring safety from *Clostridium botulinum* while avoiding hazard from nitrosamines; see page 54 for permitted limits.
- **Sulphur dioxide.** In the EU and UK this is permitted in fresh sausages and burger meat, up to 450 ppm.
- **Benzoates and parahydroxybenzoates.** In the EU and UK these are permitted in pâté, liver sausage and the jelly of meat pies and on the surface of dried meat products. In the USA they are classified as GRAS (Generally Regarded As Safe) and therefore permitted in meat products.
- **Sorbates.** These are permitted in the USA but in the EU only in certain products.

Preservative interactions

The final shelf-life of any product and its freedom from harmful organisms is determined by the interaction of all of the separate preservative and anti-preservative effects which may be present, such as:

- initial load of micro-organisms
- initial and residual concentrations of preservative
- heat process time and temperature
- salt concentration
- pH
- presence and type of polyphosphates
- storage time and temperature.

The interactions among these factors are complex and not yet fully understood. Research in the area continues, with the obvious aim that greater knowledge of these things will permit better and more refined formulation of products so as to ensure maximum microbiological safety along with improved acceptability and decreased risks of harmfulness to consumers.

One approach has been to estimate 'risk factors'. For example, it might be suggested that:

- if the input nitrite in a given product were increased from 100 ppm to 200 ppm, the risk from botulism might be reduced to 0.33 times the original (whatever that might have been);
- if the input nitrite were **decreased** from 100 ppm to 50 ppm, the risk from botulism might be increased to 3.3 times the original.

Another approach is the idea of 'hurdle' technology (Leistner, 1995). In this it is considered that salt, pH, heat process, etc., represent a series of 'hurdles' put in the path of a living microbe, which the microbe must surmount to survive. If any of the hurdles is raised high enough, the organism will not survive. However, the quantitative expression of this – how much of an increase in one 'hurdle' might be substituted for how much of a decrease in another – is far from clear in all cases.

Perhaps the most useful attempt to date is Food Micromodel, a computer program giving the growth curves of individual organisms under measured conditions of time, temperature, pH and water activity. This can be used to estimate the risks of outgrowth of harmful bacteria in foods of varied composition and storage conditions.

Heat processes

- Foods are **pasteurised** if cooked at 70°C (160°F) and held at that temperature for at least 2 minutes. At higher temperatures the holding time may be reduced. All growing microbes are killed but some spores will survive; these are unlikely to cause spoilage or poisoning in the short term but may grow to new 'vegetative' organisms if they are held at ordinary temperatures.
- Food is **sterilised** if cooked until the whole of it reaches 100°C (212°F) and is held at that temperature for several hours, or 120°C (250°F) and held for several minutes; under these conditions spores as well as vegetative organisms are destroyed.
- The shelf-life of food after pasteurising or sterilising is improved by proper cooling and protection from recontamination (as in a can). However, if re-contamination is allowed, since all living microbes have been eliminated by the heat process, the new contamination will have no competition and can grow readily. Moreover, if the new contamination contains a high proportion of harmful organisms, the food will spoil or become poisonous relatively quickly – hence the severe dangers of cross-contamination of cooked food from uncooked material.

Packaging systems

Vacuum packing

Vacuum packing is generally useful for the suppression of most of the spoilage bacteria, thus increasing shelf-life, since these bacteria require oxygen for normal growth.

Note, however, that spores are not killed but remain dormant and may cause problems after the pack is opened. For this reason it is recommended that vacuum packs for retail sale should be allowed **maximum shelf-life** of **10 days** from the date of packing (i.e. marked with appropriate 'use by' dates), **unless** some other preservative system is present, such as heat treatment or curing (Advisory Committee Report, 1992).

With **fresh (uncured and uncooked)** meats the absence of oxygen causes loss of the characteristic red colour of the meat (see page 62). Vacuum packing is therefore rarely used for retail packs, but is widely used for the long- or medium-term storage of primal cuts in the trade.

With **cured meats** the advantages are:

- Because oxygen is excluded from the packs, microbial growth is slowed down. The small amount of growth which does take place produces carbon dioxide which further slows down any continuing growth; shelf-life is therefore extended.
- Exclusion of oxygen also improves colour stability (page 72).
- Moisture loss is eliminated.
- In addition, there are the usual marketing advantages of pre-packaging: convenient units for display, sale and purchase; hygienic handling in the shop and afterwards.

With **cooked cured meat** there is the disadvantage that:

- in transparent vacuum packs the cured colour is unstable to light, especially ultraviolet light. See page 74. This problem may be overcome, however, by the use of UV-barrier packaging film.

Modified atmosphere packaging

This is a variant in which the pack is made initially with a high proportion of carbon dioxide in the headspace atmosphere, for the preservative effect of the CO_2.

Fresh meat for pre-packed retail sale may be packed with a mixture of CO_2 10–15%, oxygen 85–90%; the high oxygen content preserves the red colour of the meat.

With fresh meat to be held in bulk (e.g. primal cuts) and especially with cured meats, it is desirable to keep the oxygen concentration as low as possible, and preferable to eliminate it altogether. Atmospheres of 100% CO_2 may be used but 40–80% CO_2 is more usual, with 20–60% nitrogen. Small amounts of an inert gas such as argon are also useful, for reasons which are not yet clear.

Sous-vide

In this process vacuum packaging is combined with a mild heat treatment (e.g. 100 minutes at 70°C, 158°F, or 10 minutes at 90°C, 194°F) followed by storage and handling at carefully controlled low temperature (0–3°C, 32–37°F). It may be used for lightly cooked meats or more complex products such as ready meals, to give products without added preservative, with minimum flavour loss and commercially acceptable shelf-life up to 8 days.

Active packaging

Certain substances may be incorporated in the packaging material or in devices inserted in the package, to increase shelf-life by changing the conditions within the pack after the pack is sealed.

Oxygen scavengers are the commonest. These are small closed sachets or other containers (labelled 'Do Not Eat'!), containing reducing powders, usually iron- or ascorbic acid-based, which are able to remove the last traces of oxygen, e.g. any which may be dissolved in a cured meat product.

Canning

This is a special case in which care is taken:

- to heat the product sufficiently to destroy bacteria and all harmful spores; and
- to provide secure and robust packaging to prevent any re-contamination.

See page 112 for the canning process; page 114 for flexible pouches; page 160 for pasteurised cured meats.

FOOD POISONING

Food poisoning is usually of microbial origin. It can arise in two main ways:

- **infection**: consumer eats food containing live organisms, usually but not always in large numbers; or
- **intoxication**: consumer eats food containing toxins produced by organisms which may or may not still be present.

The symptoms include diarrhoea and/or vomiting as the body tries to get rid of the poison. There is variation among different organisms in the severity of the symptoms and in the time between consumption of food and onset of symptoms. Table 5.1 gives a summary of the main organisms responsible, their commonest origins and the symptoms they produce.

Note: Clostridia form spores, which

- are resistant to heat, so may survive heat processes which destroy the ordinary (vegetative) organism
- grow well in absence of air.

They are therefore especially dangerous in:

- food kept warm after cooking (allowing spores to germinate);
- vacuum packs

Table 5.1 Organisms causing food poisoning

Organisms	Main sources of origin	Effects
Staphylococci, bacilli	(a) Nasal passages of animals and people; therefore • people with colds, etc. • head meat of animals • other meat to a lesser degree. (b) Infected wounds, boils, etc. • people • animals – should be dealt with at meat inspection (c) Note that some staphylococci are salt-tolerant and may therefore survive in cured meats.	(Intoxication) Severity: slight to moderate Onset: rapid, 2–8 h
Salmonellae, Yersinia, Campylobacter	Intestines of animals and people: • Meat becomes contaminated at the carcass-dressing stage. • Some people may be carrying salmonellae themselves without symptoms, and may infect meat.	(Infection) Severity: moderate to severe; fatal in extreme cases (e.g. small children, old people). Onset: slow, 12–24 h
Clostridia	Soil; therefore in dirt associated with animals Spices	(Intoxication) Severity: *Cl. perfringens* moderate; *Cl. botulinum* often fatal Onset: moderate or slow, 8–24 h or longer

- large portions of produce which may be contaminated in the middle (e.g. stuffed turkey, pâté in large containers).

Accounts of the causes and control of food poisoning will be found in standard textbooks, e.g. Harrigan & Park (1991).

PRACTICAL: THE ESSENTIAL THINGS TO DO

For fresh meat and fresh meat products

(1) Start with as few microbes as possible

- Aim at **good hygiene** and **minimum delays** through slaughter, dressing, deboning and transport (see below for general principles).
- If these stages are not under the factory's own control, the supplier(s) should be known and firm specifications should be set and applied. A reasonable specification for the total viable surface count of purchased

boned-out meat would be: reject at 10^6 organisms per g or per cm^2 (1×10^4 is possible routinely under good conditions; 1×10^3 is possible with care).

- **Proper rotation and stock control** of raw materials should be applied and the factory **plant** and **premises** should also be **clean**.

(2) Minimise microbial growth during process

- **Growth is greater on cut meat surfaces**. When the meat has been cut or comminuted, it should be used with minimum delay. Meat fragments, drip, juice, etc., should not remain about – on equipment, work surfaces, or out of sight – to become sources of further contamination. Spoilage is obvious when the count reaches approximately 10^8 organisms per g or per sq cm. The absolute maximum of shelf-life is the number of days for this to take place.
- **Temperature control is critical**. Meat should be chilled or frozen when necessary. Abuse of chillers, etc., for example by overloading, poor control of doors, etc., may reduce their efficiency severely. Meat should not be kept in warm working areas for any longer than necessary.
- Long shelf-life is obtained only when both the initial count and the storage temperature are low. Shelf-life is greatly reduced by quite small increases in storage temperature in the range −1°C to +6°C (30°F to 43°F).

See the IFST *Guidelines for the Handling of Chilled Foods* (IFST, 1990) for many practical aspects of this.

(3) Do not introduce contamination

- *Factory hygiene*. Poorly cleaned parts of equipment or premises harbour microbes which can contaminate fresh products passing through.
- *Personal hygiene*. Clean protective clothing and hand washing are essential to prevent the introduction of microbes carried by staff. Their effect on morale in general, pride in the work, etc., and therefore their indirect influence on factory hygiene, may be as important.
- *Pest control*. Rodents, insects or birds may introduce direct contamination to the product; their control has good side-effects upon hygiene in general.
- *Delays in production*. When there are production delays, care should be taken with regard to the material at the bottom of the pile; it may stay there so long that it deteriorates and becomes a hazard to cleaner, faster moving material.

For cooked meat and cooked meat products

(4) Cook thoroughly to destroy all spoilage and food poisoning organisms

Cooking to temperatures at which the meat 'looks cooked' is usually sufficient, or better, use thermometers to measure core temperatures, which should be 70°C , 158°F, or higher.

In most cooking processes spores are not killed but they will present no problem so long as they are prevented from growing by carrying out the following processes properly.

(5) Cool thoroughly, in clean conditions

- Prevent delay at warm temperatures
- Food should either be eaten while still hot or be cooled rapidly and efficiently, frozen if necessary.

(6) Avoid cross-contamination

Contamination of cooked products (pasteurised or sterile) by uncooked meat, or by equipment or staff that have been in contact with uncooked meat, **must be avoided at all costs**. Because the cooked meat is 'clean', any new contamination, e.g. with food poisoning microbes, may grow particularly quickly. At least different equipment and staff should be used for the uncooked product and flow lines organised to avoid crossing over. Ideally, different rooms should be used.

Note: Supervisory staff who must move between departments present special hazards. Their hands should be washed frequently. Movements and contacts between relatively clean and relatively dirty areas (e.g. slaughter lines) should be controlled and minimised.

Cooked food

Cooked food which has gone cold *may* be reheated, provided it is *thoroughly heated*, i.e. to 70°C (158°F) minimum centre temperature, to destroy any microbes which may have grown since the first cooking, and *for immediate consumption*, to avoid any possibility of outgrowth of surviving spores. If there is any doubt on either point, the food should not be reheated.

6 Chilling and Freezing

DEFINITIONS

Distinction must always be made between:

- **Chilling** – cooling material down to temperatures just above the freezing point, e.g. +5°C (41°F) or 0–2°C (32–36°F).
- The term **superchilling** is sometimes used of chilling to temperatures just above or below the freezing point, e.g. −2 to +2°C (28–36°F).
- **Chill storage** – holding material at chill temperatures (0–5°C, 32–41°F).
- **Freezing**
  - Converting unfrozen material (usually at a chill temperature, e.g. 5°C (41°F) into frozen material; this requires a relatively very large amount of refrigeration capacity (page 96).
  - Cooling the frozen material down to its storage temperature.
- **Frozen storage** – holding frozen product at an appropriate temperature, usually −18 to −20°C (0 to −4°F); mainly achieved by keeping the environment (atmosphere) at the correct temperature.

If the wrong equipment is used for any of these stages, the correct results cannot be obtained. In particular, a cold store cannot be used to freeze large or moderate amounts of material; the result is likely to be an increase in the temperature of the store and therefore a reduction in its performance as a store.

Frozen materials of major interest to meat product manufacturers are:

- raw meat, fat, etc., intended for use as manufacturing raw material;
- finished products intended to be sold as frozen foods.

Relevant temperatures are summarised in Table 6.1.

TEMPERATURE MEASUREMENT

In measuring temperature, distinction should be made between:

- Air temperature in the equipment, plate temperature in plate freezers, liquid temperature in immersion freezers, etc. This is relatively easy to

Table 6.1 Relevant temperatures in chilling and freezing

°C	°F	
10	50	Maximum temperature specified in EU Regulations for cutting rooms, etc. Lower limit of microbial danger zone (page 00)
7	44	Growth of salmonellae ceases
0	32	Freezing point of pure water. Growth of all human pathogenic bacteria ceases
−1	30	Water in meat begins to freeze
−2	28	*c.* 50% of water frozen
−4	25	Maximum rate of deterioration of frozen meat (microbial, enzymic, chemical)
−6	21	One-star domestic freezers
−8	17	*c.* 90% of water frozen
−10	14	Parasites *Cysticercus bovis* and *Trichinella spiralis* destroyed in 2 weeks (approx.). Microbial activity ceases. Mould growth ceases
−12	10	Two-star domestic freezers
−13	9	*c.* 95% of water frozen
−18	0	Three- and four-star domestic freezers. Recommended storage temperature for general purposes
−20	−4	*c.* 98% of water frozen
−30	−22	Typical air temperature in air blast freezers. Oxidative changes in fats virtually cease. *c.* 100% of water frozen

measure using thermometers in the equipment. It is possibly not uniform throughout. It may be a very misleading guide to the temperature of material in the store.

- Surface temperature of the product. This can be readily measured using contact thermometers.
- Centre temperature of the product. This can be measured in experimental conditions:
  - insert a thermometer, probe, etc., into the product and freeze with it in place;
  - make a hole after freezing, using a drill, stainless steel spike, etc., and then insert a thermometer; allow for 1–2°C (2–5°F) temperature rise due to drilling.
- Average temperature of the product. This is the temperature which will be reached throughout the product when it is allowed to equilibrate, usually for at least several days. It is difficult to calculate beforehand.

EFFECTS OF FREEZING

The main purpose of freezing is to maintain quality for longer periods than would be possible at higher temperatures. The aspects of quality which may be affected are considered below.

In general, and under proper conditions, deterioration in quality is small or very slow, but may be serious under conditions of abuse.

Microbiological effects

Microbiological growth ceases at about −10°C (+14°F) (and growth of food-poisoning bacteria at +7°C (+44°F) but most bacteria are not destroyed even at the lowest cold storage temperatures. (Some reduction in numbers may occur but this may be ignored.)

The microbiological quality of frozen food cannot be better than it was before freezing. Therefore

- freeze only material of good quality, hygienically processed, etc.;
- freeze as soon after preparation as possible to minimise further microbial growth;
- freeze as quickly as possible.

Note that growth will resume when the material is thawed. (For special problems of the thawing stage see page 103.) After thawing, growth will continue at the same rate (related to the material temperature) as if it had never been frozen.

In chilling and freezing carcasses there may be a reduction in microbial numbers due to drying of the surface (page 97). In boxed meat (polythene lined) or properly packaged meat products, this drying does not occur.

Mould growth

Can occur at temperatures down to about −10°C (+14°F). Mould in frozen foods is a sign that either there was contamination and growth before freezing or that the cold store temperature is too high.

Until the 1950s black mould ('black spot') was not uncommon in cold stores which operated then at temperatures such as −10°C (+14°F). It usually usually about 2 months for the mould spots to become visible.

Chemical effects

Toughness in relation to time of freezing

The key factor is whether rigor mortis has commenced or been resolved at the time of freezing. (See pages 10–11 for full discussion.)

Fat rancidity

Rancidity is a major cause of stale flavours in stored frozen meat, whether cured or uncured. It is usually the factor which limits the frozen storage life.

The most important form is due to oxidation (oxidative rancidity); this is usually measured by peroxide value (PV). Lipolytic rancidity, which tends to cause soapy flavours and is measured by acid value or free fatty acid (FFA), also occurs but is less significant for present purposes. For fuller consideration consult the appropriate chemistry textbooks.

Factors which accelerate the development of rancidity include:

- **Air or oxygen**. Vacuum-packed meats, if the packaging is perfect, will be trouble free. Perfect packaging is rare.
- **Light** speeds up the chemical changes. In chemical terms 'free radicals' are encouraged.
- **Softer fat**, in chemical terms more unsaturated fat, e.g. pork, is more prone to rancidity than beef (see page 18).
- **Meat pigments** or **blood pigments**. In chemical terms, fat oxidation and pigment oxidation are autocatalytic: each process encourages the other. Pigments may be in contact with the fat in comminuted products or where hygiene is poor and fat surfaces are smeared with blood, meat juice or meat fragments.
- **Frozen conditions**. There are two effects:
  - (i) chemical changes go more slowly at lower temperatures, and
  - (ii) as ice is formed, reacting substances are concentrated in the remaining water, thus making the chemical changes go faster. In the case of rancidity, this effect is greater than the low-temperature effect.

  The worst temperatures for rancidity are about −2 to −4°C (28–25°F), where the temperature is not very low but a high proportion of the water is frozen.
- **Salt**. This may accelerate the chemical changes, possibly because it reduces the availability of water, therefore operates in the same way as solid ice (above).

Because of the link with changes in the meat pigments, development of rancidity is commonly associated with deterioration of colour. For example, use of frozen fat which has begun to oxidise can cause rapid discoloration in hamburgers.

Pastry

There are no special problems with unbaked pastry. Baked pastry may stale rapidly at thawing temperatures (page 180).

Physical effects

Ice formation – mechanical and chemical effects

When meat is frozen, it is the water which freezes, forming crystals of ice. The location of the water in unfrozen lean meat is shown in Fig. 2.1. The shape and size of the ice crystals formed may affect other properties of the meat. The quantity of ice present in the meat is constant at any given temperature but the size of the crystals depends on the rate of freezing.

- **Fast freezing**. Many small crystals are formed simultaneously, inside and outside the cells; water bound to myofibrils is the last to freeze (Fig. 6.1(a)).
- **Slow freezing**. First, crystals form in the water outside the cells. These then grow as liquid water is attracted on to them. The result is large ice crystals, mainly outside the cells, and some dehydration of the cell contents (Fig. 6.1(b)).
- **Fluctuating temperatures in frozen material**. When the temperature rises some ice is melted. Normally the ice is lost equally from all of the existing crystals so that the smallest crystals may disappear. When the temperature falls, ice is re-formed on the remaining crystals, causing them to grow.
- **Freezing in fatty tissue**. The water present in the connective tissue in fatty tissue freezes in a similar way. There is no dehydration of the cells on slow freezing but dissolved substances become concentrated in the unfrozen water, as with lean meat.

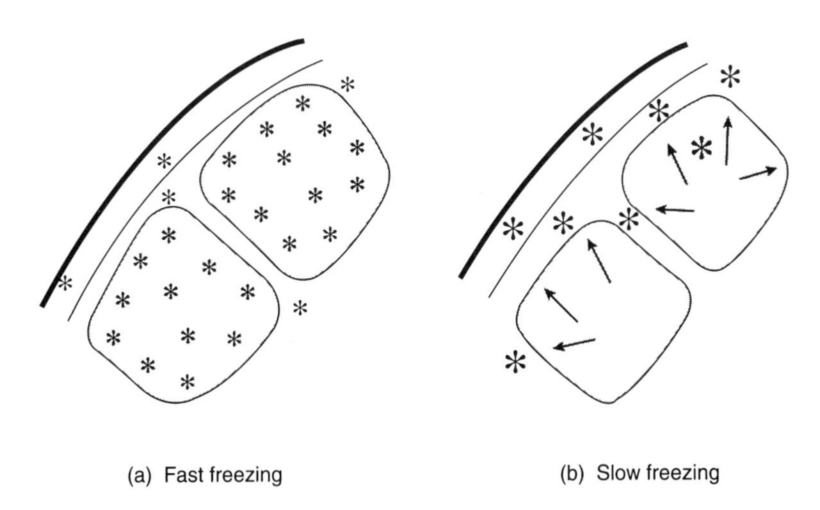

Fig. 6.1 Ice formation in meat.

Heat considerations

Heat must be extracted from meat as it is cooled and frozen, or supplied when the frozen meat is thawed. For pure water the relevant quantities are shown in Table 6.2.

Table 6.2 Heat considerations in freezing and thawing

Cooling	Heating	Amount of heat to be extracted (cooling) or supplied (heating) kcal/kg	kj/kg
Cooling from body temperature (37°C, 98°F) to 0°C (32°F)	Heating from 0°C (32°F) to 37°C (98°F)	37	155
Freezing water at 0°C (32°F) to ice at 0°C (32°F)	Thawing ice at 0°C (32°F) to water at 0°C (32°F)	80	335
Cooling from 0°C (32°F) to −20°C (−4°F)	Heating from −20°C (−4°F) to 0°C (32°F)	10	42

The largest single factor is the conversion of water to ice (or vice versa) with no change in temperature (extraction of the 'latent heat'). Figure 6.2 shows typical cooling curves for meat frozen in an air blast at −30°C (−22°F). The flat section (A) represents the long time needed to freeze the water in the middle of the meat. Note the general conclusions:

- The main burden on a freezing plant is to freeze the meat; cooling the frozen meat from freezing point to −20°C (−4°F), for example, is relatively easy.
- Conversely, any failure in the refrigeration plant, or serious overloading, etc., is likely to mean that some of the meat does not get frozen at all.

Evaporation

The atmosphere in a freezer or cold store is normally very dry. This is because:

- The freezing coils or plates are the coldest part of the unit;
- Water therefore condenses on the coils or plates: 'frosting';
- Water vapour is therefore drawn to the atmosphere round the coils from the atmosphere further away, i.e. the atmosphere further away becomes drier.

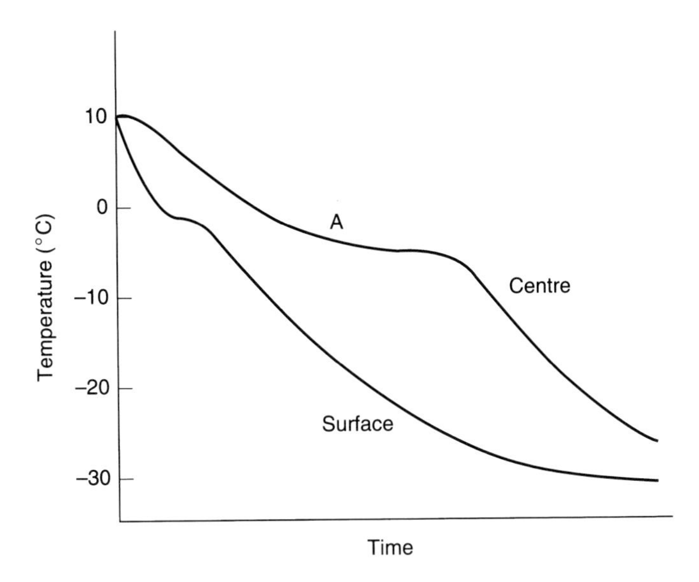

Fig. 6.2 Freezing a large piece of meat.

Therefore any materials in the store which contain water or ice and are not protected against moisture loss will lose moisture to the atmosphere of the store. The materials will dry out and the moister atmosphere will contribute to further frosting of the coils. The loss of water from the material will appear as a weight loss.

Freezer burn

Freezer burn is the result of sublimation of ice from unprotected surfaces of meat, etc. In the early stages there are small white spots or greyish areas – sometimes called 'freezer scorch'. This discoloration may disappear when the meat thaws out and the meat surface rehydrates. If the freezer burn is severe, it will not disappear on thawing. It is harmless but the appearance, eating quality and manufacturing properties of the affected meat are adversely affected. When freezer burn occurs, there is likely also to be significant weight loss.

Freezer burn and the associated loss of weight can be avoided if the following points are noted:

- Prevent drying out by effective packaging – polythene or other moisture-proof material.
- Prevent damage to packaging by careful handling.
- Still-air stores are better than blown-air stores.

EFFECTS ON MEAT PROPERTIES

Colour

Red meats (beef, mutton and pork) are darker, more brown/grey when frozen than when unfrozen.

Poultry dark meat, rabbit and game birds may have a purple appearance when frozen, redder when unfrozen.

Poultry white meat is red-brown if frozen slowly, cream to white if frozen rapidly. These colours revert to cream on thawing. Dark colours are usually taken as a sign that poultry meat has been thawed and re-frozen slowly through malpractice.

Drip losses

It is now known that the drip losses from meat which is frozen and then thawed, and consequently the potential losses when the meat is cooked, are greatly affected by the **rate** of freezing. The faster the meat is frozen, the lower the losses. For the best results, attention should therefore be paid to:

- the size of the pieces to be frozen – the smaller, the faster freezing;
- the operating temperature and efficiency of the freezer – the colder the faster.

Cooking losses

Lean meat

- Water losses are only slightly increased by freezing and thawing.
- The fat binding capacity of the meat, which depends on the lean meat–water–salt system, is only slightly affected by freezing and thawing of the lean meat, except that some cuts of beef forequarter may be affected. The reasons for this are not yet clear.
- Lean meat may be comminuted in the frozen state without further effect on the lean meat–water–salt system and the fat binding capacity but note that in the frozen state the water is not available to dissolve salt or protein; the water–salt part of the system is effective only in the thawed state. The reduction in freezing point caused by the salt is not sufficient to counteract this.

Fatty tissue

- Fat losses are not increased by freezing and thawing alone, i.e. if the fat is fully thawed and used at the same temperature as if it had never been frozen.

- However, fat losses are increased at lower temperatures of **comminution**; fat comminuted when frozen always gives very high losses (page 36).

Storage life

Some typical values for storage life are given in Table 6.3.

MANAGEMENT OF THE COLD CHAIN

Chillers, refrigerators

These are devices for maintaining an air space at temperatures close to the freezing point of water. They operate by the alternate compression and expansion of a refrigerant gas. There are three essential parts:

- **Compressor**: the gas is compressed by a pump and is heated in the process.
- **Condenser**: the gas passes through pipes or plates cooled on the outside by air or water or a combination of both; the cooling causes the pressurised gas to liquefy.
- **Evaporator** or **cooling coils**, located in the space to be cooled: the pressure is released, the liquefied gas vaporises, drawing heat from the surroundings in order to do so.

Refrigerants

For many years the non-toxic CFCs (fully chlorinated and fluorinated hydrocarbons, with trade names Freon or Arcton) were preferred, until the discovery of the damaging effects of gaseous chlorine compounds on the ozone layer. Now, under international agreement (the Montreal Protocol and EC Regulation 3093/94), the use of CFCs has been phased out and HCFCs (partially chlorinated and fluorinated hydrocarbons) are to be abandoned by 2020 or 2030. HFCs, which contain no chlorine, remain permitted.

Ammonia, once being displaced by CFCs, is now returning to widespread use. It is effective and economical but highly poisonous. (The characteristic smell is a safeguard in case of small leaks, but ammonia must be considered a hazardous material.)

Butane is also coming into use but here the problem is its flammability.

Freezers

The term ‘freezer’ is used here to indicate equipment for freezing foods, etc., not cold stores for holding foods.

Table 6.3 Storage life of frozen meats (from *Recommendations for the Processing and Handling of Frozen Foods.* International Institute of Refrigeration, 1964)

	Temperature		Expected storage life (months)
	°C	°F	
Beef	−12	10	5–8
Roasts, steaks, packaged	−15	5	6–9
	−18	0	8–12
	−24	−10	18
	−18	0	12
Ground meat, packaged (unsalted)	−12	10	5–6
	−18	10	4–8
Veal			
Roasts, chops	−18	0	8–10
Cutlets, cubes	−18	0	6–8
Lamb	−12	10	3–6
Roasts, chops	−20 to −18	−4 to 0	6–10
	−23 to −18	−10 to 0	8–10
	−18	0	12
Pork	−12	10	2
Roasts, chops	−18	0	4–6
	−23	−10	8–10
	−29	−20	12–14
	−18	0	6–8
Ground, sausages	−18	0	3–14
Pork or ham, smoked	−18	0	5–7
Ham, fresh	−23 to −18	−10 to 0	6–8
Bacon, fresh (green)	−23 to −18	−10 to 0	4–6
Poultry, eviscerated, in moisture-proof wrapping	−12	10	3
	−18		6–8
	−23 to −20	−10 to −5	9–10
Fried chicken	−18	0	3–4
	−29	−20	14
Rabbit	−23 to −20	−10 to −5	Up to 6
Offals, edible (packaged)	−18	0	3–14
Lard	−18	0	9–12

Design capacity

Note that every freezer was designed to freeze:

- a specified volume, size and shape of product
- at a specified throughput rate
- from a specified input temperature.

Any increase in the throughput rate or input temperature will probably mean that the product will be frozen less completely: its surface may appear as normal but less of the centre will be frozen and it will equilibrate to a higher average temperature. There is likely therefore to be additional cooling load on the cold-store equipment when the product is moved there.

Types of freezers

Blast

This is the most common type:

- Air blast at −30 to −40°C (−20 to −40°F).
- Air speed 5 m per sec (1000 ft per min).
- Batch or continuous operations.
- Will handle many shapes and sizes.

Plate

- Comprises refrigerated heavy metal plates; product is pressed between them.
- Used for uniform rectangular packs or blocks.

Liquid immersion

- Used for poultry, especially turkeys.
- Packaged produce is immersed in a tank or spray of refrigerant, normally calcium chloride or propylene glycol solution.
- Gives a very fast 'crust' freeze.
- There are problems with leakage of refrigerant into packs; this is not harmful but may necessitate washing and re-freezing.

Liquefied gas

- Liquefied non-toxic gas is sprayed in the vicinity of the product on a conveyor, etc; freezing occurs as the gas is vaporised. **Nitrogen** boils at −196°C (−321°F). **Carbon dioxide** sublimes at −78°C (−108°F).
- The process is very rapid.
- Excellent for small articles, e.g. chicken portions.
- With large articles there are problems of surface cracking when the centre expands on freezing after the crust has set.
- Relatively expensive; should be used only when the advantages of very rapid freezing are required.

Frozen stores

Temperature problems

Stores are basically designed to keep the air cold and the assumption is that the product comes in already fully frozen. Note abuse problems:

- Moist air in the store leads to frosting of cooling coils, etc.; even warmer air in the store; more frequent defrosting.
- Failure to carry out the freezing stage properly (above), which puts an extra burden on the store refrigeration, therefore failure to maintain store temperature properly.
- Stacking should allow movement of air among boxes of the product.

Freezer burn (page 96)

Evaporation from the freezer-burned surfaces may also lead to extra frosting of cooling coils and problems as above.

Illumination

Note colour and rancidity problems in frozen meat products (pages 66, 74, 94).

The 'greenhouse effect' sometimes occurs in cold stores and often in retail display cabinets. If food in transparent containers (e.g. over-wrapped trays) is placed in the light, radiation from the light raises the product temperature without changing the external air temperature (as in a greenhouse). Quality deterioration from this cause can be rapid.

Frozen transport

Problems are mostly those of temperature maintenance. In principle they are similar to those of frozen stores, but exaggerated:

- Opening the doors, switching off the cold unit, high product input temperatures due to delays during loading, etc., all have larger effects because the chamber is relatively small with little reserve of cold.
- Insulation is likely to be less than in a store.

THAWING

Heat transfer from a warmer source

Thawing by heat transfer from a warmer source is a fundamentally less efficient process than freezing. Getting heat into the product depends on:

- **Thermal conductivity**. The thermal conductivity of water is much smaller than that of ice. Thawing is therefore a slower process than freezing because heat has to pass through an increasing thickness of water instead of an increasing thickness of ice, as shown in Fig. 6.3.
- Temperature difference between the product and the source of heat. This is usually determined by the temperature of the outside source. A large temperature difference can be arranged for freezing but it is usually difficult to arrange a high external temperature for thawing because of microbiological problems (see below).
- Nature of the heating medium. Water transfers heat more efficiently than air but also causes more hygiene problems (see below).

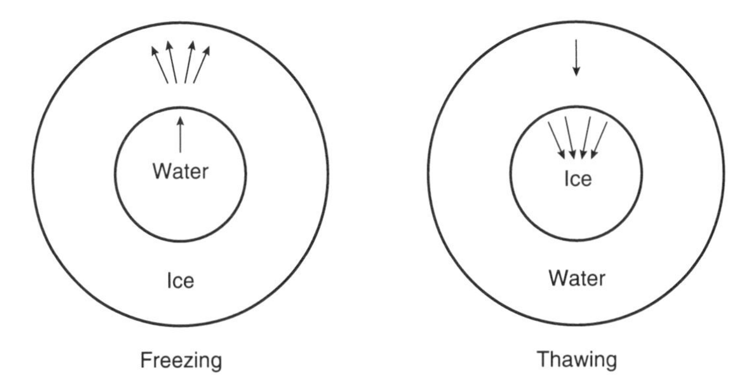

Fig. 6.3 Thermal conductivity during freezing and thawing.

Conditions for microbial growth are much better during thawing than during freezing because:

- There is moisture and nutrient for microbes on the outside of the meat as soon as thawing commences; if water is used, this may also be a medium for growth.
- The higher the external temperature the faster the rate of microbial growth; the lower the thawing temperature the slower the thawing and the more time for microbial growth; the best compromise is probably to use air at 5–7°C (40–45°F), gently circulated with a fan.

Generation of heat within the product

Thawing by this method can be rapid and avoids most of the microbial problems.

- Microwave (radio frequency heating): it is good for units of uniform size and shape. There may be problems of overheating or 'runaway thawing' in some parts of the product.

- Dielectric heating is also good for units of uniform size and shape.

Both require expensive equipment; therefore a large throughput is needed to justify their use.

Re-freezing

For various reasons given above, thawed or partly thawed meat should be regarded as fresh meat with a life before spoilage of at most a few days. Provided this is recognised and allowed for, such thawed meat can technically be re-frozen if this is done properly. In practice, however, it may be difficult to ensure that the whole operation is done properly and that proper allowance is made for the reduced shelf-life of the final material.

Reasons for *not* re-freezing therefore include:

- The fact that spoilage has already commenced in the thawing stage is disguised; on thawing out the second time, the storage life is already that much shorter.
- Re-freezing, unless properly done in a blast freezer, plate freezer, etc., imposes a burden on the refrigeration equipment for which it was not designed.

7 Cooking

Almost all meat products are cooked either by the manufacturer or by the consumer, and sometimes by both. Exceptions are:

- raw hams
- dried, fermented sausage
- dried meat such as biltong, jerky, pemmican (none of these three is common in the UK)
- underdone parts of steak, etc. (see below).

ADVANTAGES OF COOKING

Primary advantages

Meat is made more palatable.

- It is made more tender. With pure muscle (e.g. prime steaks), the effect is quite small because the meat is already quite tender, therefore minimum cooking is required, e.g. rare steak, underdone roast. Less 'noble' meat, i.e. most of the meat in manufactured products, must be cooked to soften connective tissue.
- The colour changes: pink colour in cooked ham and fried bacon (page 71), and brown colours in roasts, grills, etc. (page 68).
- Flavours and aromas are developed (page 77).

Harmful bacteria, parasites etc. are killed. Cooked meat is therefore safer to eat than uncooked meat. See Chapter 5 for details. Note especially the requirement to avoid recontamination after cooking.

Exceptions, as noted above, are:

- Underdone cooked meats. These are safe where
  - the underdone part is in the middle of an uncut piece of meat (e.g. steak, roast), and therefore not contaminated with bacteria, etc.; or
  - the meat is treated carefully and served very soon after cutting up (e.g. steak tartare), so that any contaminating bacteria, etc., do not get time to grow;
  - in the case of well aged beef, the brown 'cooked' colour does not

appear until the temperature is well over the 70–72°C (158–162°F) necessary for microbiological safety. (But note the possible problem of 'premature browning', p. 68.)

- Dried sausages, raw hams, etc. In these the control of harmful bacteria results from the composition of the product: salt, acid, nitrite and especially low moisture content (low water activity).

Secondary effects

- Comminuted or fabricated products bind together on heating.
- Cured colour is 'fixed' or converted to the denatured form.
- Pastry of pies, etc., is baked.
- Effects on nutritional value:
  - **Proteins** are denatured; this has little detrimental effect on nutritional value except that over-cooking can reduce the available lysine content.
  - **Vitamins and minerals**: meat is an important source of iron and certain vitamins in the average diet. Except for vitamin B_1 they are relatively unaffected by cooking; see Table 7.1.

Table 7.1 Nutrients in meat and effects of cooking

Nutrient	% Contribution to average intake	Effect of cooking
Iron	38	None
B_1 (thiamin)	22	Large reduction
B_2 (riboflavin)	30	Small or no reduction
Nicotinic acid	41	Small or no reduction
A	32	Small or no reduction
D	39	Small or no reduction

CHANGES WHICH OCCUR ON COOKING

The effects to be described below are summarised in Fig. 7.1. Note especially the following.

Water boils at 100°C (212°F)

(The boiling point in meat is actually slightly higher than 100°C because of the effect of dissolved substances; however, the difference is slight and we will ignore it here.)

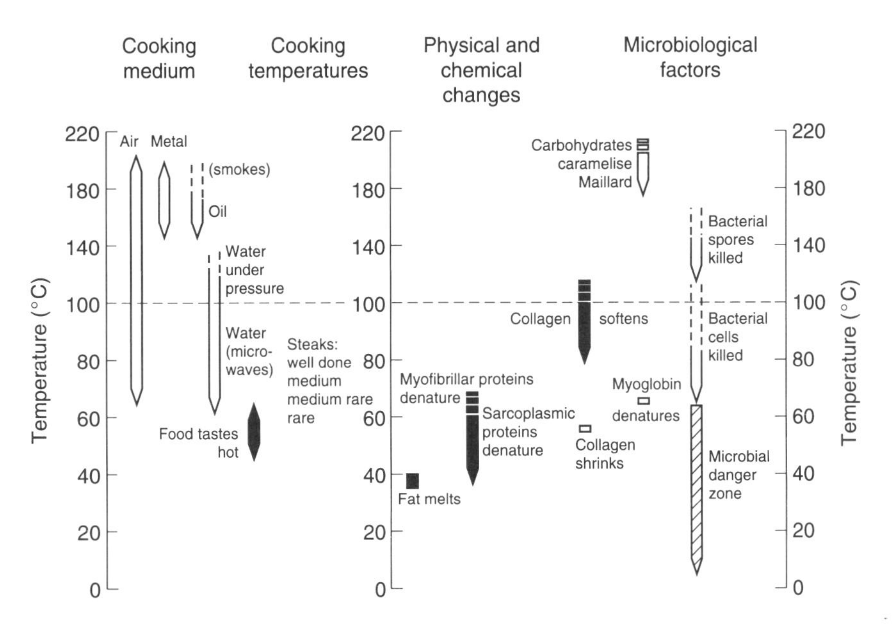

Fig. 7.1 Significant temperatures in cooking.

- So long as water is present, the temperature cannot be higher than 100°C; because of the high water content of lean meat, this means that in practice the temperature in the middle of almost all pieces of meat will not exceed 100°C, whatever the external cooking conditions.
- Cooking under pressure raises the boiling point of the water so that all of the meat can reach higher temperatures, up to 120–125°C (248–257°F); this occurs in canning and retort cooking.
- Only if the outer surface can dry will it be able to reach higher temperatures; this occurs in roasting, grilling and frying.

Protein changes

- **Sarcoplasmic** (soluble) **proteins** are progressively precipitated from about 40°C (104°F). Thus meat becomes paler because of increased light scattering (page 00). The majority of the precipitation occurs above 60°C (140°F); it is complete by about 70°C (158°F).
- **Contractile** (myofibrillar) **proteins** – actin and myosin – are denatured at about 65–70°C (149–158°F). **Myosin** exudate heat-sets at this temperature.
- **Connective tissue** (collagen) shrinks at 55–60°C (131–140°F); (therefore there is an increase in water losses at about this temperature). (See page 000 for significance in sausage bursting.) Collagen softens at about 80–100°C (176–212°F) in the presence of water and hydrolyses to gelatin

from about 90°C (194°F) upwards. Ideally, 25–40% of collagen should be hydrolysed; if there is more than this the texture may be adversely affected. Without added water it becomes hard and dry and in the presence of water it becomes soft and disintegrated).

- **Meat colours** (myoglobin, nitrosyl myoglobin) denature at 65–75°C (149–167°F):
  uncured meat: red → black (page 68)
  cured meat: red → pink (page 71)

Cooking losses

Water loss

The extent of water or fluid cooking losses depends mainly on the product temperature and to a smaller degree on cooking time. There is a large increases in loss in the temperature range 50–60°C (122–140°F); 80–100% of the total loss has occurred by the time the sample reaches 80°C (176°F).

In a large piece of meat or meat product, overall losses depend on the temperatures reached at different depths. So, for example, for the same product cooked to the same centre temperature:

- rapid cooking gives high surface temperature, therefore high losses at the surface, therefore high total losses;
- slower cooking requires lower surface temperature (same centre temperature), therefore lower losses at the surface, same losses at the centre, and lower total losses.

Some of the water lost may be re-absorbed if the meat remains in contact with the liquid during cooling.

'Sealing' of meat surfaces by 'searing' is sometimes said to reduce overall losses but there appears to be no evidence that this is so.

Fat loss

Fat melts at 37–40°C (98–104°F). Free fat may therefore escape from a product mixture at quite low temperatures unless held in an effective matrix. Fatty tissues remain relatively undamaged up to 130–180°C (266–356°F), although some individual cells may burst at lower temperatures, e.g. 50–55°C (122–131°F).

Changes above 100°C (212°F)

Dark colours are formed by:

- dehydration
- charring (over *c.* 150°C, 302°F)

- Maillard-type browning reactions (over *c.* 150°C, 302°F). These changes require amino acid, especially lysine, alanine from meat protein; plus reducing sugars e.g. glucose, lactose from added sugar, honey, milk powder, etc.); or oxidising fats (therefore encouraged by fat, air).
  Note that these changes cannot occur below 100°C (212°F); the temperature in the centre of the meat cannot exceed 100°C (212°F) as long as any water is present, so the above changes can occur only at the surface.

Flavour changes

The chemistry of the flavour of meat is very complex and not well understood in detail. A number of general points, however, are fairly clear:

- The flavour of raw meat is not very strong.
- It is somewhat metallic, due to the iron in myoglobin and haemoglobin.
- The characteristic flavours of different meat species appear to reside mainly in the respective fats.
- Hard fats such as beef and mutton have high proportions of solid material at the temperature of the human mouth; therefore they form solid, tallowy deposits which most people consider unpleasant; softer fats such as pork fat do not do this.
- The older the animal, the stronger the flavour. Thus beef flavour is stronger than veal, laying hen is stronger than broiler chicken, etc.

Pleasant and interesting flavours are developed on cooking. The main effects are as follows:

- Where much connective tissue is present in the lean meat, this is hydrolysed and softened (page 20).
- There is an increased sensation of juiciness; therefore whatever flavours are present are more readily perceived.
- Breakdown of some meat constituents gives inosine, a degradation product of ATP, ADP, etc., which has a characteristic meaty flavour, and glutamic acid by protein hydrolysis, which enhances flavours by sensitising the palate. (*Note:* monosodium glutamate (MSG) is widely added to foods for this purpose; 0.1% is usually sufficient.)
- Many other flavoursome substances are produced, mostly in trace amounts.
- Strong flavours are produced by the caramelisation and Maillard reactions at temperatures of 150°C (302°F) upwards, i.e. at the meat surfaces during roasting, frying, etc.
- Strong flavours and aromas characteristic of the meat species are produced by heating fatty tissues in air; thus dry rendered chicken fat, beef dripping, etc., are good sources of 'chicken', 'beef', etc. flavour.

- Added flavours from spices, herbs, etc., are blended and 'rounded' during the cooking process.

Note also that:

(a) The flavour of fried bacon consists of

- the taste of salt
- the aroma of heated pork fat
- a small but significant flavour developed from nitrite with lean meat.

(b) Unpleasant flavours may be due to:

- **Boar taint**: a flavour in the meat of older boars due to relatively high levels of the sex hormone androsterone; it is more readily detected and more disliked by women than men; it is said to be below the level of detection in young boars (page 000). Ram taint may occur in older sheep but is very uncommon.
- **Catty taint**: an aroma or taste of tom-cats, produced by reaction between sulphides (from meat proteins) and unsaturated ketones such as mesityl oxide (which may occur in paint thinners and many other products).
- **Other taints**: usually due to contamination by flavoured foreign materials.

Microbial danger zone

At temperatures between 10°C and 63°C (50–145°F), most of the microbes which cause spoilage and all those which cause food poisoning can grow rapidly. This temperature range must be avoided as far as practicable. Therefore, for instance, the UK Food Hygiene Regulations require food to be held for the minimum practicable time at temperatures between 10°C (50°F) and 62.7°C (145°F) and state that they should be rapidly heated to, or cooled to, a higher or lower temperature.

Note that the colour change from raw to cooked occurs at 65–75°C (149–167°F) (see above); thus meat which *looks* cooked may be taken to *be* cooked from the microbiological point of view.

PRACTICAL COOKING CONDITIONS

Source and effectiveness of heat

The amount of heat transferred into the meat during cooking depends on:

- the temperature difference between the meat surface and the cooking medium, i.e. on the **cooking temperature**
- the rate of input of heat to the surface, i.e. the **heat transfer coefficient** of the heating medium
- the **total cooking time**.

For the conditions normally encountered in cooking meat, these factors are summarised in Table 7.2.

Table 7.2 Meat cooking mthods and temperatures

Cooking medium	Cooking temperature	Heat transfer
Dry air (roasting, baking)	Up to 200°C (392°F)	Fair
Water (stewing, braising)	Up to 100°C (212°F)	Good
Water under pressure	100–125°C (212–257°F)	Good
Moist or saturated air (steaming)	Up to 100°C (212°F)	Very good, because of latent heat from condensing steam
Moist or saturated air under pressure (retorting)	100–125°C (212–257°F)	Very good, because of latent heat from condensing steam
Fat or oil (frying)	150–190°C (302–374°F)	Very good
Metal	Up to 200°C (392°F)	Good at points of contact, otherwise as air
Radiation		
grilling	Depends on conditions	Very good
microwave	Temperature reached depends upon the energy input; is not usually above 100°C, 212°F	Very good

Heating conditions within the product

The deep layers in the meat or meat product are heated from the layers above. They have high water content, i.e. conditions within the meat are similar to those of heating in water whatever the heating conditions at the surface; the internal temperature does not rise above 100°C, 212°F (approx), unless heating under pressure.

The rate of heating depends on:

- thermal conductivity
- surface temperature.

The method of heating does not otherwise have much effect. In microwave heating the effects are similar to water cooking, unless drying out occurs, e.g. at corners – 'hot spots'.

SMOKING

Originally hams, etc. were hung in fireplaces or chimneys to dry out. Smoke gives better keeping quality and characteristic flavour due to:

- reduced moisture content
- preservative action of some of the smoke constituents (phenols, etc.)
- antioxidant action of some of the smoke constituents
- flavour of the smoke constituents.

For more details see page 151.

CANNING AND HEAT PROCESSING

This may be regarded as a special type of cooking process, in which the amount of cooking is controlled to ensure long-term stability by inactivating all micro-organisms which could cause spoilage or food poisoning.

As commonly understood, the term covers rigid hermetically sealed metal containers (cans), but the principles may be extended to other kinds of containers – jars, bottles, flexible pouches, sealed cartons, etc.

The essential features are:

- prevention of spoilage and minimising of contamination before processing;
- establishment of the correct heating conditions (the 'scheduled process');
- consistent application of the scheduled process in manufacturing practice; maintaining integrity of the container and minimising post-process contamination.

Heat process values

The theory of can processing starts with the need to provide bacterial safety by preventing the growth of *Clostridium botulinum*. The spores of this organism are the most heat-resistant poisoning organism known, and the vegetative organism if it grows out from the spores may be the cause of severe illness with a high fatality rate.

The amount of heating required is usually expressed in terms of an F_0 value:

$1 \times F_0$ = the amount of destruction of *Cl. botulinum* spores which occurs in 1 minute at 121°C, 250°F (or equivalent times at other temperatures).

The agreed amount of destruction to give safe protection from *Cl. botulinum* is a heat process equivalent to:

$F_0 = 3$

This is called the **minimum botulinum cook**.

Exceptions, where the heat process value may be different from $F_0 = 3$, are as follows:

- High acid foods with pH below 4.5. Here the food is preserved and kept safe by the acidity, and a 'botulinum cook' is not necessary. Meat products do not usually come into this category.
- Most (uncured) meat products and very many other foods receive heat processes in excess of $F_0 = 3$ (e.g. $F_0 = 8$ or higher) in order to control spoilage organisms which are more heat resistant than *Cl. botulinum*, e.g. flat sour organisms, which produce acid and turn the product sour without producing gas, so do not 'blow' the can but leave it flat.
- Cured meats receive heat processes below $F_0 = 3$ because of the preservative effect of nitrite in controlling *Cl. botulinum*. Recommended minimum conditions are:

  meat of good microbial quality (clostridial spores not more than 1 per g)

minimum content of sodium nitrite	75 ppm input
minimum salt content	3.5% on moisture
minimum heat process	$F_0 = 0.1$

- Large hams intended to be stored under refrigeration may even lower heat processes, not measurable as F_0 values. The limiting conditions have not been defined.

In practice, the required process value is obtained by operating a **scheduled process**, in which the operating times, temperatures and pressures of the retort in which the cans are cooked are closely specified for the retort operator. The details will be different for different products; they may differ even for different retorts or filling lines in the same factory. The scheduled process must be followed exactly.

(Details of the measurements and calculations necessary to establish a scheduled process will be found in standard texts, e.g. Stumbo (1973), Hersom & Hulland (1980), and Larousse & Brown (1997).)

Special cases

HTST (high-temperature, short-time) and UHT (ultra-high temperature) processes

These processes use the fact that heat-induced quality deterioration may be much reduced if a required F_0 value is delivered at a higher than usual temperature and therefore for a shorter than usual time. They are used, for example, in milk processing and may be used for some ready meals, but are of little benefit for the majority of heat-processed meat products.

Aseptic canning

Aseptic canning can be defined as the filling of a sterilised product into a previously sterilised container, immediately followed by hermetic sealing with a sterile closure in an atmosphere free of micro-organisms. By these means the total heating of the food product can be minimised without sacrificing shelf-life or safety. The technology is complex and normally requires long runs of product in a fluid form, adapted to rapid heating and cooling, e.g. in plate heat exchangers. This process is little used for meat products.

Sterilisation in flexible pouches

Advantages

- Pouches are flat and relatively thin, so heating and cooling can be fairly rapid; therefore some of the improvements in quality of HTST processing apply.
- Flexible packs are relatively easily opened and need no can opener.
- The pouch material is claimed to be inert; therefore there is no contamination of the contents.

Disadvantages

- Pouches are not very robust.
- Pouches are made of multilayer material for maximum strength; therefore they are expensive; they may de-laminate.
- An external carton is always required for protection.
- Heat seals cover a large area and are soft after heat process; considerable care is needed during the cooling stage and immediately after processing; very gentle handling is needed to prevent micro-leakage and therefore contamination.

Flexible pouches are therefore useful mostly for high value or high quality products, e.g. some pasta products where over-cooking would be detrimental to the texture.

To overcome the dangers, suppliers of pouches lay down Codes of Practice covering processing, handling and packaging conditions.

Corned beef

Although corned beef normally contains sufficient nitrite to preserve it, it is given a full sterilising heat process in order to cook to the extent required for flavour and texture.

Part Two
APPLICATIONS

8 Comminuted Meat Products

Recall (page 7) that many comminuted meat products originate as solutions to the problem of how to make palatable products from:

- less 'noble' cuts of meat, i.e. those with high connective tissue content dispersed through the meat, or high fat content dispersed through the meat;
- meat trimmings produced in preparing more 'noble' cuts for retail sale or for manufacturing into bacon, ham, etc.;
- fat removed from various other cuts to upgrade them.

Comminution of these raw materials has the following useful effects:

- pieces of meat of mixed sizes and shapes are brought to a uniform, more attractive condition;
- connective tissues are broken up and rendered less obtrusive and more easily softened on cooking;
- fat and lean can be minced together so that large or moderate proportions of fat become less obtrusive;
- the comminuted lean meat binds the whole mixture together, especially when salt is present;
- the texture and eating quality of the products are, therefore, different from those of the starting materials, and more desirable to consumers.

Variations in starting materials and processing details yield a wide range of products, to be considered below.

COMMINUTION PROCESSES

Dicing, etc.

Meat in dice, chunks, strips, etc. may be presented in sauces, gravies, etc., or as a component of some complex sausages. The process is easily done on the small scale with a sharp knife. In the past it was difficult to produce neat dice, etc. on production machinery, especially with very soft meat such as chicken, mainly because of the difficulty of cutting cleanly through tough connective tissue and soft lean meat at the same time. However, there are now several satisfactory machines for this purpose. See Fig. 8.1.

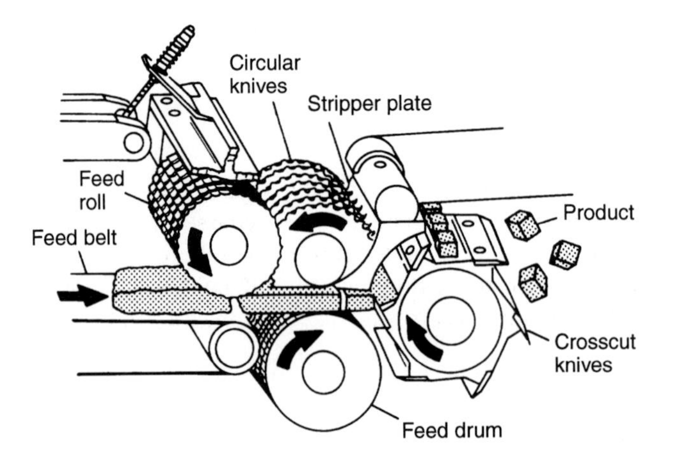

Fig. 8.1 Urschel dicer/strip cutter.

Flaking

Impeller flaker

See, for example: the Urschel 'Comitrol'. Meat pieces enter at A, are thrown by the impeller B against the sharp edges C, which are arranged in a static ring; flakes exit at D. See Fig. 8.2. This method cuts slivers of meat of constant thickness and parallel sides. Thickness is adjustable by changing the ring of knives C. Connective tissue is cut very cleanly.

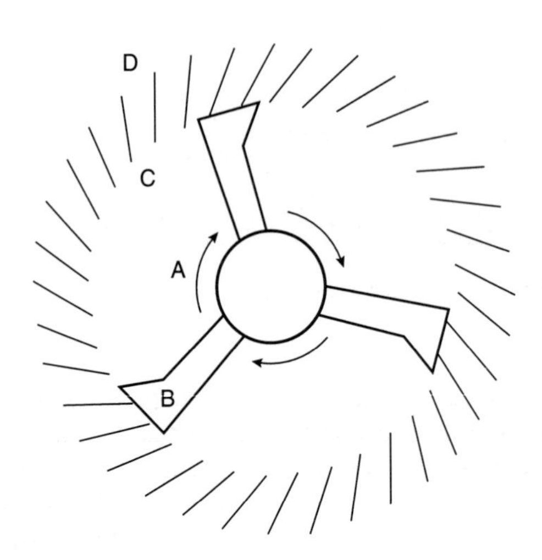

Fig. 8.2 Flaker (diagrammatic).

Frozen or unfrozen meat may be used, pre-broken into appropriate sized pieces (200–500 g depending on machine size).

Block flaker

In the Bettcher block flake, for example, a guillotine cuts flakes from the end of a frozen block of meat. The flakes are coarser than those given by the Comitrol. Optimum flaking temperature is −2 to −4°C (28–24°F); if warmer, there is tearing without cutting; if colder, the flakes crumble. Temperature normally rises by only *c.* 1°C (2°F) during flaking.

Chopping

Rotary bowl chopper

There are many makes of rotary bowl chopper, also known as bowl cutters or silent cutters. All are similar in principle (see Fig. 8.3). A set of three or more (up to 12) curved knives rotates at high speed in a vertical plane close to the surface of one side of a curved bowl which itself rotates slowly in a horizontal plane. Various shapes of knife are available. Care and some skill are required to sharpen, set and maintain the knives.

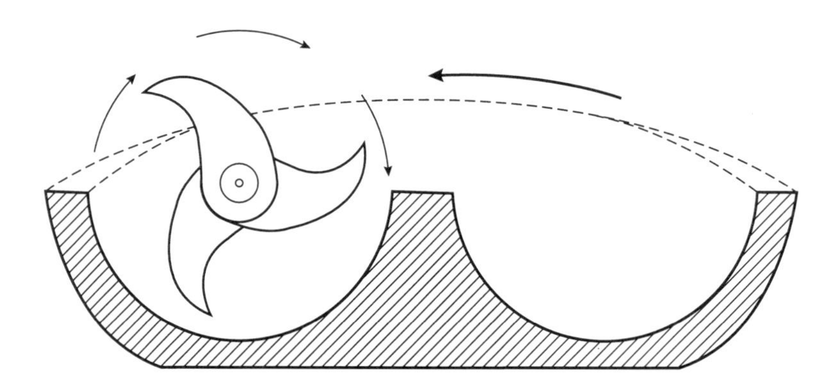

Fig. 8.3 Rotary bowl chopper (diagrammatic).

Meat in the bowl can be cut very finely, depending mainly on residence time. In addition to the vigorous cutting action, the massaging effect of the sides of the knives on the mass of chopped meat may be important. Satisfactory chopping temperatures range from −1°C (30°F) (initial) to +22°C (70°F) (final). Colder temperatures lead to damage to knives; warmer temperatures lead to overchopping of fatty tissue and release of free fat.

Some modern machines are provided with a tilting mechanism to assist unloading. Some also have close-fitting lids and exhaust pumps to enable chopping under vacuum.

Stationary bowl chopper

In these machines the cutting action is provided by rotating knives at the bottom of a stationary bowl. Mixing is ensured by a scraper or mixer blade driven from above. A dosing funnel, tilting gear for emptying, steam or gas injection, an outer jacket for heating or cooling and vacuum facilities may all be provided as in Fig. 8.4.

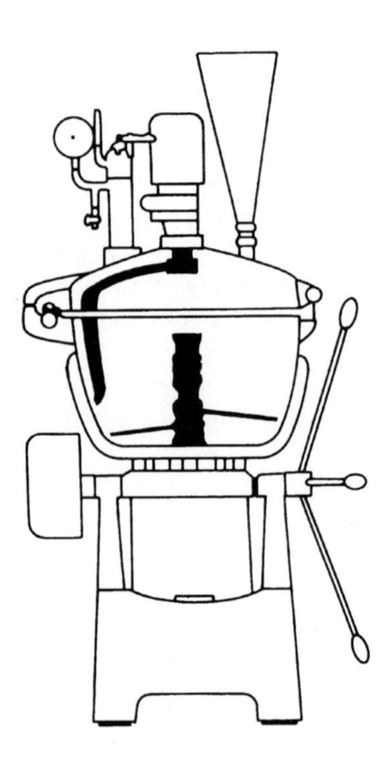

Fig. 8.4 Stationary bowl chopper.

Mincing (grinding)

There are many commercial makes of mincer. The principles are illustrated in Fig. 8.5.

Note, however, that:

- Considerable pressure is put on the meat in the screw feed chamber.
- Tearing occurs between the screw flights and the chamber wall.
- Final comminution occurs when portions of meat extruded through the rotating inner plate of knives are either passed on through holes in the fixed outer plate or sheared off as the holes pass out of register.

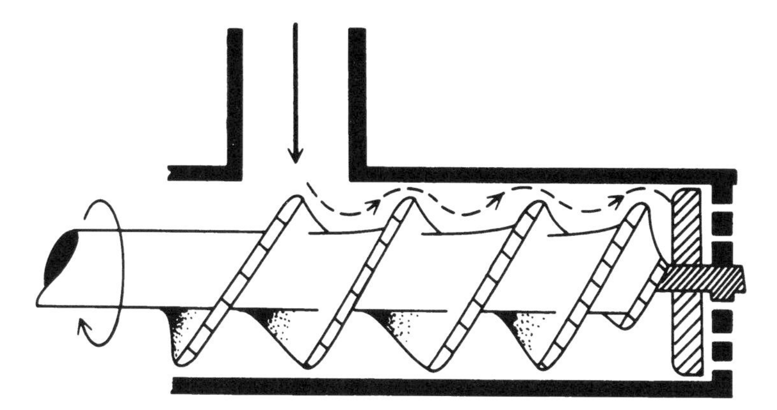

Fig. 8.5 Mincer (diagrammatic).

There is thus great tearing, pressure and shearing and the meat is not cut cleanly or finely. Connective tissue is divided fairly well in a sharp mincer but may be troublesome in a blunt one. For damaging effects on fatty tissue, see page 00. Sharpening of mincers is somewhat difficult, especially, for example, in the cases of the edges of the screw flights and the faces of the plates. Mincing is usually done on unfrozen meat, so no melting effect (latent heat) is produced to absorb heat. The temperature may rise by up to 10°C (18°F), especially with small hole sizes. See page 35 for effects of mincing temperature on fat losses.

Milling

For examples are the Karl Schnell (KS) Mill, Mincemaster, Lowboy, etc. See Fig. 8.6. This type is similar to a mincer, except that there is:

- no feed screw: the plates are mounted horizontally and fed by the weight of material above;
- a rotating knife which moves at much higher speed;
- some cutting or tearing action in the space between the plates as well as at the edges of the holes in them.

There is finer comminution than in a mincer and the operation is much faster.

Because of the fineness of comminution and intimate mixing, when used with fatty materials these machines may be referred to as emulsifiers.

Mechanically recovered meat (MRM or MSM)

This not strictly a comminution process, but the product is delivered in a very finely comminuted form (page 12). It is not normally convenient to add other substances while the comminution is taking place, so some of the

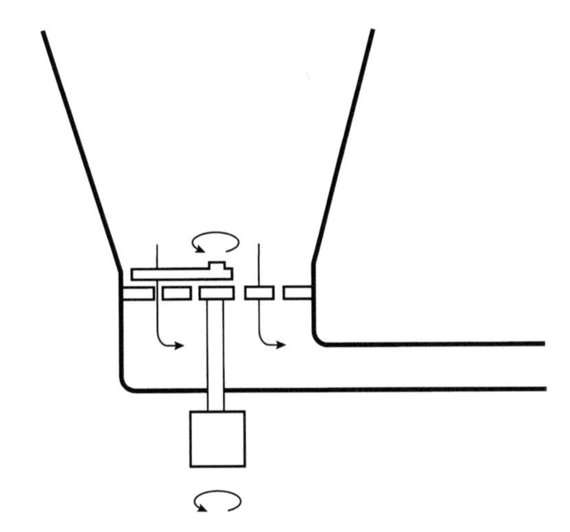

Fig. 8.6 Mill (diagrammatic).

advantages of mixing (see below) are not obtainable until after the MRM is made, when the cohesiveness of MRM may make mixing more difficult.

Mixing with other ingredients

Comminuted meat products contain ingredients other than comminuted meat. It is often convenient to be able to incorporate these during the comminution process. Bowl chopping and milling permit this very readily.

Where ingredients can be added during the comminution process, there may be technical advantages. Note especially questions relating to

- dispersion of salt into lean meat;
- mechanical working of the meat in the presence of salt (page 33).

Other comminution processes do not readily permit such additions and must be followed by a further stage in a suitable mixing machine.

Mixers also provide mechanical action and the effects mentioned above, but less efficiently, since the lean meat–salt interactions can take place only after the mixing process has dispersed the salt into the meat.

PRODUCTS

Burgers, meat balls, re-formed meat: definitions

A **hamburger** was originally a large all-beef sausage of a kind made in Hamburg, usually served in slices which would be fried or grilled. It is now commonly made, especially in North America, in the form of flat patties.

In the UK the names '**burger**' and 'beefburger' have become common, partly through the belief that a hamburger might be made with ham. 'Baconburger', 'lamburger', etc. are names given to products made with bacon, lamb, etc.

In the UK also, the names 'hamburger', 'burger', 'beefburger', etc., may refer to a meat-with-cereal product with 80% minimum meat content, although products with higher meat content, up to 100%, are widely made also. In the UK regulations, not more than 40% of the meat content may be fat; in practice the proportion may be much lower.

Meat balls are similar products formed into small balls.

Reformed meats are products made by comminuting meat and 're-forming' it to a structure and shape (e.g. cubes) resembling unaltered meat. They are intended usually to present meat with high connective tissue content in a more attractive form.

Burgers

Principles

In products with relatively high meat content the basic problems are:

- to comminute the meat sufficiently to tenderise it by breaking up connective tissue;
- to use salt and mechanical action just enough to bind the product together, before and after cooking;

The process of hamburger making may be represented as in Fig. 8.7. Note, however, that:

- excessive comminution, salt and/or mechanical action give very firm binding and cohesion, but also loss of fibrousness and increase in rubbery texture (page 44) (in this context, mechanical action includes all movement during the manufacturing processes – mixing, stirring, flow in pipes or through pumps, action of forming machines, etc.;
- where the fat content is high (say over 20%), the need for fat binding may raise further problems (page 132);
- where lean meat content is low, because of high fat content and/or low total meat content, these problems are magnified because there is less lean meat to provide the essential binding of the product.

Comminution

This can be done by:

- Flaking, using an impeller flaker or block flaker (pages 120, 121);
- Mincing (page 122); a 'mince–mix–mince' process can also be used;

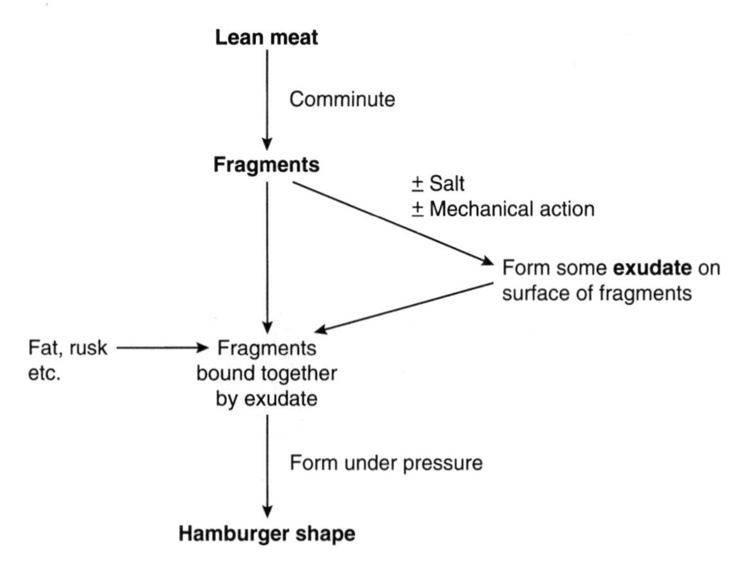

Fig. 8.7 Hamburger process.

- chopping tends to give a too highly comminuted product, especially if high quality (i.e. non-disintegrated texture) is required.

Forming

There are several methods of forming, as shown below.

- **Moulding, hand pressing**. This is a simple process, adequate for many purposes. It is slow and does not give high pressures. See Fig. 8.8.
- **Extrusion and slicing**. The material is formed into a tube or log and cut into slices.

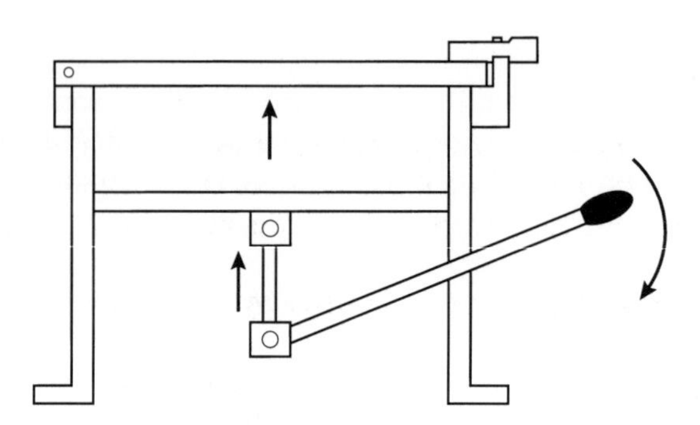

Fig. 8.8 Hand-operated patty press (diagrammatic).

- **Extrusion moulding**. See, for example, the Hollymatic patty former shown in Fig. 8.9. In this process, high speeds are possible and pressures are fairly high. Note the shearing forces across the face of the burger as it is separated from the residual material in the feed system.

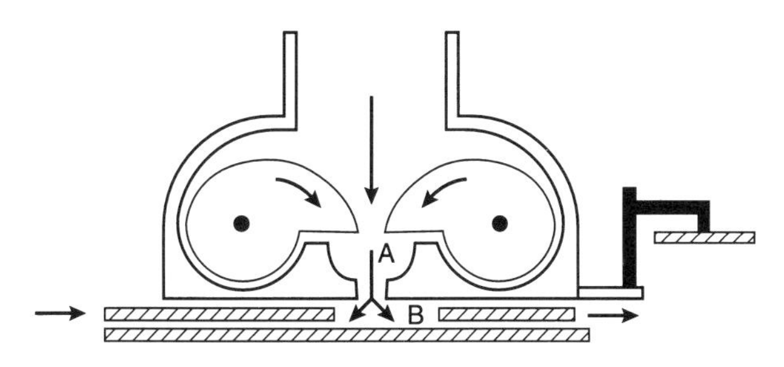

Meat in chamber A is pressed into patty form in chamber B

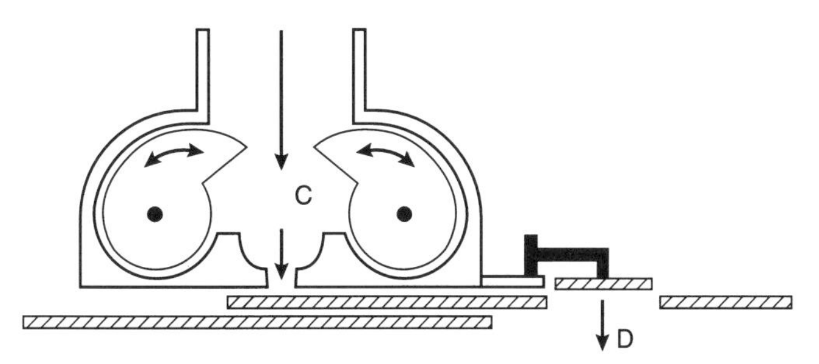

Fresh charge of meat enters at C; patty previously formed is discharged at D. Separating paper may be supplied at this point.

Fig. 8.9 Hollymatic 2000 sliding plate patty former (diagrammatic).

Meat binding

The problem is usually to obtain sufficient exudate formation for good binding without too much loss of texture from the fragments being bound, which leads to a rubbery product. A 'mince–mix–mince' process is fairly satisfactory.

Note that mechanical action is provided by:

- Handling operations before forming (e.g. pumping, stirring, etc.). 'Setting up' may also occur, causing mixtures to become more sticky and less able to flow.

- Forming machines, especially extrusion moulding. Shearing forces, leading to alignment of meat fibres, may cause the finished burger to shrink along the fibre alignment when cooked, which distorts a circular shape (overcome by forming elliptical burgers which become circular on cooking).

Temperature effects

During comminution	see pages 121, 123
During mixing	temperature rise is normally small, unless mixing is prolonged
During forming	usually negligible
During general handling	NB problems caused by undue delay

Flavour

Problems may be associated with deterioration of fat, especially in frozen storage (before or after incorporation into the product). See page 94 and below under 'Colour'.

Variations in spice flavour, onion flavour, etc., may be due to incomplete mixing.

Colour

In beef products particularly, loss of redness and change to brown or grey is a common problem (pages 62ff). Note particularly the ill effects of the following:

- *High temperature*, which accelerates all other effects.
- *Undue microbial growth*, indicating poor initial quality of meat or other ingredients; (note that onions, fresh or dried, may be an important source of microbial contamination; overcome by pre-cooking the onion); delays during handling, processing, etc., may also be responsible.
- Enzyme activity in meat (greater in fresh meat).
- Oxidising agents, e.g. cleaning materials. Also, the pungent principle in onions (pyruvic acid) can cause brown discoloration (possible cures, precook onion by blanching, frying, etc; use less onion; use less pungent onion). In addition, fats with actual or incipient rancidity can cause colour problems, which are usually evident before there is any flavour problem. See also page 94.
- Incomplete mixing, especially of MRM, which may be more highly coloured than other ingredients and also more difficult to mix.

Re-formed meat

Properties

As usually understood, a re-formed meat should have the following properties:

(a) Texture, colour and flavour similar to those of good quality lean meat.
(b) For commercial reasons it will usually contain a proportion of less expensive meat, including

- forequarter meat, trimmings, etc., with high connective tissue content
- rind, sinews, etc.
- MRM.

(c) Products sold in the raw state, to be cooked by the consumer, should have cooking properties as close as possible to those of steaks, roasts, etc.
(d) Or the product may be cooked during production to resemble a meat product normally sold in the cooked state, e.g. ham, diced steak in a sauce or gravy.

Re-forming processes

Some meat products are made by pieces of meat being bound into larger units, usually by exudate formation and heat setting, e.g. restructured ham, cured picnics, and various related ham-like products; also some beef roasts.

Flaking and compaction is the most common process for making steak-like products for subsequent cooking. Flaking cuts connective tissue into thin slices, so that large pieces will not be encountered on chewing. Being thin they may also cook more readily, becoming further softened. The best process is to:

- flake at *c.* −2°C (28°F)
- hold the flaked meat for several hours with occasional stirring
- form under pressure into logs or patties of the required shape
- freeze.

Under these conditions good binding occurs without the use of salt; the flavour of the product is, therefore, more steak-like. The reason for the good binding in the absence of salt is not clearly understood.

Structured pastes

Structured pastes used in cooked meat products, e.g. pies, canned stews, etc., are made by the following process.

Comminuted meat is intensively worked with salt and water to give a flowable paste (bowl chopper, mince–mix process, KS mill, etc.), then extruded and set (e.g. by extrusion into hot water or into a chamber heated by microwaves). The extruded rod can then be cut into dice, etc.

Other materials may be mixed with the paste to give discontinuity and complexity of structure, e.g. pieces of meat, fatty tissue, etc.; fibres prepared from cooked meat; fibres or pieces of vegetable protein or mycoprotein.

Cooking

Care must be taken that the cooking process is done thoroughly in order to ensure proper pasteurisation. The centre temperature must reach not less than 70°C (158°F) for 2 minutes. After cooking, the product must be chilled thoroughly and hygienically to avoid any recontamination or bacterial growth under unnecessarily warm conditions. For uncured meats it is recommended that the centre temperature be brought to 5°C (41°F) in not more than $8\frac{1}{2}$ hours

Meat analogues

These of course are not meat products but may be mentioned here in passing. They are made mainly from non-meat proteins, either vegetable (soya, etc.) or cultivated mould mycelia (mycoprotein). Most are available as dried chunks. The texture when rehydrated may be a more or less passable imitation of that of lean meat.

They may be used, with proper declaration on the label, as ingredients in the cheaper kinds of meat products, to provide additional bulk or texture. Their protein content may not, of course, be counted towards the meat content.

Pet foods

Note that the principles and methods of making re-formed meats and meat analogues are extensively used in making 'chunky' pet foods.

Sausages

The primary economic purpose of these products is (or was originally) to present relatively large proportions of fat in palatable ways. Comminution of the fat is therefore a common feature.

Definitions

Sausages are defined in the UK by their shape, 'approximately cylindrical, with hemispherical ends'.

In this handbook we distinguish three broad types; among the sausages made in various countries there are many variations in detail and much overlapping among the types.

Fresh sausage

- Uncured and uncooked, therefore short shelf-life
- Various degrees of comminution, but most products tend to be coarse
- The British type of fresh sausage can have various levels of meat content.

Emulsion sausages

- Made from finely comminuted lean meat and fat
- Often further processed, e.g. by cooking, addition of curing salts, smoking, drying; may therefore have intermediate shelf-life
- Some contain quite large inclusions of coarsely cut meat, fat, spices, etc.

Fermented dried sausages

- Long shelf-life (e.g. 1–2 years) obtained by
  - lactic acid produced by fermentation at the beginning of processing
  - curing with nitrite, usually formed microbiologically from nitrate
  - drying out in the later stages of manufacture.
- The process is complex and difficult to manage.

Liver sausage and similar spreadable products are considered with pâtés later in this chapter.

Fresh sausages

Principles

(For meat binding, see pages 38–46; for fat binding see pages 28–35.)

Table 8.1 summarises the chopping conditions required when sausage making.

These considerations lead to the scheme shown in Fig. 8.10 as the ideal process for an average fresh sausage, using separated lean meat and fat.

Practical formulation and processing

Table 8.2 shows characteristic properties of a range of sausage ingredients.

Table 8.1 Chopping conditions for sausage making

A. The best and worst conditions for chopping lean meat or fat:

	Best conditions	Worst conditions
For **lean meat binding** (and finer texture)	high salt concentration moderate added water phosphate long chopping time communication at chill temperature	no salt no added water poor distribution short chopping time
For **fat binding**		
• retaining cellular fat	short chopping time moderate added water comminution when warm	long chopping time dry chopping mincing or flaking comminution when frozen
• binding free fat	as for lean meat binding	as for lean meat binding

B. Conflicts between the 'best conditions' for both lean and fat:

Lean meat binding requires:	high salt concentration *and* moderate added water
Best compromise:	add dry salt at the beginning of chopping allow a brief period of relatively high salt concentration then add water later, gradually so that salt concentration is reduced slowly
Retention of cellular fat requires:	short chopping time
Lean meat binding and binding of free fat require:	long chopping time (therefore fine texture)
Best compromise:	**for coarse-cut products** – add fat after lean has been chopped for a short time with water; thus there is a shorter chopping time for the fat mixture, under fairly fluid conditions **for fine-chopped products** – long chopping of lean; accept damage to fat cells but bind the liberated fat in a strong lean–meat–salt matrix

Role of lean meat

Note that the lean meat, as modified by its treatment with salt, water and mechanical action is responsible for:

- binding water in the product; therefore giving low water losses and low shrinkage on cooking;
- holding fatty tissue cells and free fat in the mixture; therefore giving low fat losses;
- binding the product together.

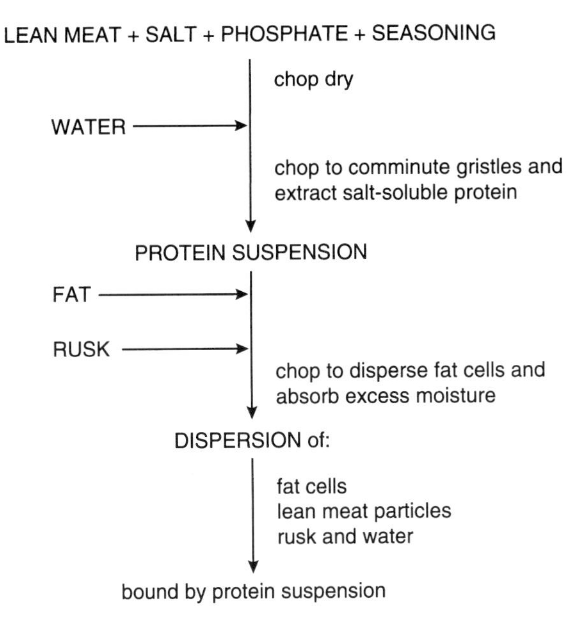

Fig. 8.10 Ideal sausage chopping process.

Fat and water losses are also connected with bursting of sausage casings on cooking (pages 139, 140).

All these quality factors therefore depend greatly on the proportion of lean meat in the recipe. The more lean meat, the better the final product and the easier it is to make.

Published tables of 'binding index' or 'binding property' (e.g. Table 8.1) can be seen to set out meats in order of their lean meat (muscle) content, with only a few special adjustments. Attempts to use more direct measures of binding index in, for instance, computer-based least-cost recipes, have not yet proved very successful.

The proportion of lean meat is controlled by the following factors:

(a) Total meat content. UK legislation provides:

- frankfurter, Vienna sausage, salami – not less than 75% meat content
- pork sausage – not less than 65% meat content
- other designations – not less than 50% meat content.

(b) Proportion of fat. UK legislation: lean meat must be not less than 50% of the minimum meat content. The lowest case, therefore, is 50% meat content with lean meat constituting 25% of the total.

(c) Inclusion of rind. In the UK, rind is counted as 'meat' in calculating the 'meat content'. However, in a declaration of ingredients only the rind

Table 8.2 Characteristics of sausage ingredients (from Long *et al.* (1982), pages 35, 36.)

Carcass part	Level of fat (%)	Colour rating[1]	Binding property[1]	Protein content (%)	Ratio of moisture to protein	Added[2] water (%)
Pork						
Blade meat	8	80	95	19.2	3.76	16.2
Liver	8	80	–	20.6	3.47	23.1
Jaw meat	8	80	80	20.9	3.40	24.9
Ears	10	10	20	22.5	3.00	36.0
95% trimmings	10	70	90	18.9	3.73	16.8
Picnic hearts	12	80	90	18.6	3.73	16.6
Stomachs, scalded	13	20	5	16.7	4.20	7.3
Spleens	15	60	–	15.9	4.33	5.2
Nose meat	15	45	70	17.9	3.74	16.2
Cheeks, trimmed	15	65	75	17.8	3.79	15.2
Weasand meat	17	80	80	16.4	4.05	10.1
Giblets	17	75	75	16.9	3.91	12.4
Hearts	17	85	30	15.3	4.40	4.2
Boneless ham	19	60	80	16.9	3.80	15.2
Tongues	19	15	20	16.3	3.95	11.9
Partially defatted chopped pork	21	50	50	17.4	3.54	19.9
80% trimmings	25	50	80	15.8	3.72	16.0
Head meat	25	50	80	16.1	3.60	11.8
Picnic trimmings	25	60	80	15.6	3.80	14.4
Neckbone trimmings	25	60	70	15.9	3.55	19.0
Skirts	30	50	45	14.2	3.90	12.6
Lips	31	5	10	20.1	3.41	23.8
Skin	32	5	20	28.3	1.40	92.6
Tongue trimmings	32	15	10	15.6	4.34	5.1
Snouts	35	5	10	14.6	3.45	19.9
Partially defatted tissue	35	15	20	14.0	3.63	16.8
50% trimmings	55	35	20	9.7	3.64	15.0
Head skin trimmings	55	15	50	9.2	3.90	12.0
Regular trimmings	60	30	35	8.4	3.77	13.1
Backfat trimmings	62	25	15	8.1	3.71	13.6
Skinned jowls	70	20	5	6.3	3.72	13.1
Belly trimmings	70	20	30	6.3	3.75	12.8

Bacon ends	70	10	5	8.8	2.40	26.5
Backfat, untrimmed	80	20	30	4.2	3.83	11.9
Beef						
Weasand meat	6	75	80	17.8	4.20	7.1
Bull meat	8	100	100	20.8	3.40	24.8
Liver	9	80	–	20.7	3.40	23.6
Imported cow meat	10	95	100	19.0	3.65	18.6
Boneless chucks	10	85	85	19.5	3.57	20.4
Tripe	11	5	10	12.8	5.90	(16.2)[3]
Melts	12	95	20	16.9	4.20	7.2
Lungs	12	75	5	17.5	4.00	11.0
Shank meat	12	90	80	16.8	4.20	7.2
Domestic cow meat	12	95	100	18.8	3.65	18.4
85/90 trimmings	15	90	85	18.9	3.45	22.7
Cheeks	15	90	85	18.3	3.59	19.4
Tongues	20	25	20	15.5	4.15	8.4
Tongue trimmings	40	15	15	12.6	3.75	14.6
Lips	20	5	20	15.9	4.00	11.0
Partially defatted chopped beef	20	50	45	20.0	3.00	33.2
Hearts	21	90	30	14.9	4.30	6.0
Head meat	25	60	85	16.4	3.54	19.4
Partially defatted tissue	25	30	25	18.9	3.20	15.2
75/85 trimmings	25	85	80	16.9	3.41	22.2
Boneless navels	52	65	55	10.5	3.55	16.3
Boneless flanks	55	55	50	9.9	3.54	16.2
Cooler trimmings	65	20	15	8.0	3.40	16.3
Beef fat	85	10	5	3.3	3.55	11.3
Beef clods	10	95	100	20.0	3.50	19.9
Veal						
Trimmings	10	70	80	19.4	3.62	19.3
Mutton						
Boneless mutton	15	85	85	18.1	3.70	17.1

1 Based upon 100.
2 Water which may be added and still comply with the US requirement that moisture in finished product shall not exceed four times % of meat protein plus 10% of finished weight.
3 Water must be subtracted.

'naturally associated' with the meat actually used may be included as 'pork meat' counted towards the 'meat content'. The proportion acceptable is not clearly established but many authorities accept the convention that total pork meat content may include 10% of rind. Any excess over this proportion must be separately declared as 'Rind'. Rind as commercially prepared may carry a variable proportion of adhering fat. For formulation purposes it should be regarded as 'Rind' (without any adhering fat) plus the visible trimmable fat In calculating the meat content the fat portion should be treated as 'fat', the remainder as 'lean'. The lowest case therefore, becomes: 50% meat content (as above), of which

- 25% (50% of 50%) is fat (including any visible fat on the rind)
- 5% (10% of 50%) is added rind, less trimmable fat
- 20% is lean meat.

Apart from questions of legality, any reduction of the lean meat content below this level will make it more difficult to produce a satisfactory product. (Added proteins such as soya isolate may help, but often not enough.)

Other factors

In practice the main modifications which may be required to the scheme in Fig. 8.10, page 133 are as follows.

(1) **Use of semi-lean meat** (pork belly, beef flank, mixed meat trimmings, etc.). The use of semi-lean meat is almost inevitable in practice; it means that it is impossible to treat lean meat and fat separately: they must be added together. Therefore:

   - Add any fat which is not in the form of semi-lean meat as late as possible.
   - If the recipe contains any lean meat (over, say, 75% VL), start by chopping this dry with salt and seasoning, then add the semi-lean meats and start adding water immediately. This will give some degree of wet-chopping conditions for the fat (page 37). If there is only semi-lean meat, start adding water as soon as possible after chopping commences.

(2) **Use of ice**. It is common to use some of the recipe water in the form of ice, to minimise the temperature rise during chopping, thereby improving hygiene, shelf-life, etc. But care should be taken not to use too much ice. Usually 50% of the water in this form is sufficient. This is 15–20% of the total recipe. When melted, it will prevent a temperature rise of about 1.5°C (3°F), which would be a normal rise in a

coarse sausage chopping. If excess ice is used, some may remain unmelted at the end of chopping with the following disadvantages:

- Solid ice will not dissolve salt; when the pieces of ice melt later they will give regions in the sausage low in salt content, therefore poor in water binding and meat binding.
- Chopping with hard pieces of solid ice may increase damage to fatty tissue and therefore increase fat losses.

(3) **Treatment of rusk**. As used in the UK and elsewhere, rusk is a specially prepared material made from wheat flour, chemically raised, baked, ground and supplied in specified ranges of particle size. It absorbs three to four times its weight of water, with a slight increase in temperature.

The rusk may be pre-soaked before use, then allowed to cool to offset the temperature rise. However, the water used for pre-soaking is probably more usefully employed in the lean meat–salt–water chopping stage or in providing fluid conditions for fat chopping (see above); on balance, therefore, it is preferable not to pre-soak.

Also, because one of the purposes of rusk is to soak up any free water in the recipe, its addition should, if possible, be delayed until after the fat has been chopped.

The addition of dry rusk at an early stage may increase the damage to fatty tissue by the abrasive effect of the hard particles before they become soaked.

(4) **Soya isolate or concentrate, sodium caseinate, etc**. Numerous tests have shown that these materials reduce cooking losses in sausages. They are normally most effective when the lean meat content is low. Use up to 3%, normally adding at the beginning of chopping. Note also the use of these materials in pre-formed emulsions (paragraph 6 below).

(5) **Rind** may be added at the beginning of chopping. It is usually better to pre-cook or pre-blanch the rind (short treatment in hot water) to soften the texture. Dried rind may also be used.

(6) **Fat emulsion or rind emulsion**. See pages 39–40 for manufacture, advantages, etc., of such emulsions. They can be added at the beginning of chopping. Note that water used in making the emulsion is not available as free water in the recipe. See paragraphs 1, 2 and 3 above.

Casings

About 1% of the sausage weight consists of casing.

Natural casings

These are made from cleaned intestines. They are usually packed with solid salt; in this state shelf-life at 5°C (41°F) is almost indefinite. Before use, shake off excess salt, and soak in cold or tepid water for about 2 hours.

The use of natural casings has diminished greatly in recent years. Disadvantages are the need to remove salt and to spool before use; variable diameter; tendency to produce curved sausages; casings are liable to damage.

Table 8.3 Quantities and dimensions of natural sausage casings (data from Gerrard, 1977)

	'Rounds' or 'runners' small intestine		'Middles', 'bungs' large intestine	
	Length per animal (m)	Diameter (mm)	Length per animal (m)	Diameter (mm)
Cattle	36–40	36–46	9–12	45–60
Sheep	22–47	18–26	5–6	*
Pig	17–19	32–42	4–5	40–45

* Not usually used for casings.

Artificial casings

(a) Regenerated collagen

These are manufactured from hides or similar collagenous material by dissolving them in acid and extruding into concentrated salt solutions (e.g. ammonium sulphate) to precipitate the protein in a continuous tube. Additives may include cellulose, cellulose derivatives, glycerol, etc. The tube is then dried, and shirred ready for spooling.

Regenerated casings are more convenient in use than natural casings; they are straight and of constant diameter. Special varieties are made for dried sausage, to adhere to the sausage during drying.

(b) Cellulose

This type is used for frankfurters and other skinless sausages. The filled sausages are passed through hot water or moist air at 55–70°C (130–160°F) to coagulate the meat surface and give some cooking throughout. The casing is then cut longitudinally, peeled off and discarded; the thin surface layer of cooked meat serves instead of a casing to hold the sausage intact.

The cellulose casing may be coloured to transfer a water-soluble colour to the sausage surface.

(c) Co-extruded collagen

This patented process is used in some Unilever factories. Sausage meat is extruded through a filler tube simultaneously with an annulus of a collagen suspension around it; the collagen sets to a firm casing around the finished sausage.

Filling and linking

Filling machines have a piston feed to a nozzle with casings fitted over the nozzle. They vary from small and hand-driven to large and fully automatic. Linking may be by hand or (in the larger machines) automatic.

Problems encountered in filling

- **Creaming**: Passage of sausage meat along filling pipes, etc. causes mechanical action in the meat next to the pipe wall, with formation of a layer of emulsified fat in a 'cream' of the worked meat. This causes pale, uniform colour and obscures any coarseness of texture or colour in the body of the sausage. The only cure appears to be a slower, gentler, less complex filling procedure, which is often not feasible.
- **Setting-up**: Undisturbed sausage mixture increases in viscosity, and becomes stiffer and more difficult to fill. This problem therefore arises if material is held over break times (avoidable) or during machine breakdowns (difficult to avoid). Stirring (if possible) reduces the viscosity increase but encourages creaming.

Bursting of sausages

When a sausage is cooked, the casing (natural or synthetic collagen) has a strong tendency to shrink at temperatures close to or just below the temperature at which the meat mixture heat-sets (page 00) and binds to the casing. In a well made sausage the casing therefore draws tight and becomes firmly bound to the meat. Pricking the casing normally makes no difference to this process. Defects can arise, however, giving sausages with different kinds of burst appearance as follows.

Cooking-out

Meat appears to have extruded from the ends of the sausage; actually, the sausage length is unchanged but the casing shrinks inwards. Immediate causes are:

- poor binding between skin and sausage meat
- lubricant (e.g. fat) between skin and sausage

therefore the intact skin can slide inwards over the sausage.

The basic cause is defects in recipe or manufacture which lead to high fat loss; the exuded fat provides the lubricant and prevents the casing from binding to the sausage meat.

Split skin

The casing splits, almost always longitudinally, and becomes more or less detached from the sausage. In extreme cases it falls off completely. Immediate causes are:

- poor binding between casing and meat
- no lubricant layer
- any damage, or undue strain, on the casing before or during cooking (e.g. pricking, intense local heating such as at line of contact with a baking tray, bubbles of steam inside the casing produced suddenly as in deep fat frying).

The basic causes here are faults in the recipes or manufacture leading to poor meat binding; rapid cooking; and cooking from frozen.

Split sausage ('butterfly', 'kipper' etc.)

The casing splits longitudinally but remains bound to the sausage meat. It shrinks laterally, pulling the meat with it and therefore tearing the sausage open. Immediate causes are

- splitting of the casing
- good binding of casing to meat.

The full explanation of this behaviour is not yet clear.

Cooking losses

Fat and water loss on cooking are commonly measured in R & D or quality control work as indices of product quality. The cooking method chosen should be appropriate to the likely market for the sausages.

The commercial significance of such results is not always clear, but physical shrinkage resulting from high cooking losses is likely to lead to customer complaints.

Solid or precipitated matter may sometimes appear in the cook-out, especially if rind or sodium caseinate has been used in the recipe. Farina (potato starch) may be a useful corrective.

Flavour

The main components of the sausage flavour are those of the meat and fat, plus salt. Herbs and some spices may be added, depending on local taste;

they do not significantly affect any other property of the sausage, except appearance in some cases.

Possible sources of flavour problems include:

- spoiled or 'sour' meat – always a microbiological problem at some stage
- spoiled or rancid fat – may be microbiological; more likely to be age, especially if frozen fat was used (page 94).
- some additives introduce flavours:

blood, blood plasma, etc.	metallic or liver-like taste at 10% whole plasma, 2–3% dried plasma or less
phosphates	metallic, bitter taste over 0.3–0.5%
soya flours	'beany' taste over, say, 3–5%

Colour

Various problems can arise.

- **Pressure marks**. These are areas of discoloration at the point of contact between tightly packed sausages. Oxygen is used by enzymes in the meat or by fat becoming rancid, or by bacterial growth; access of air from outside the pack is prevented by the close contact of the sausages; thus the pigment is converted to the purple reduced myoglobin form. In its early stages this effect is reversible: on exposure to air the pigment re-oxygenates. However, as storage proceeds, the small amount of oxygen which does penetrate between the sausages causes oxidation to the brown metmyoglobin form; at this stage the discoloration is irreversible.
- **White spot**. This defect has occurred in the past but does not now appear to be common. It is characterised by the appearance on the surface of the sausage of circular grey–white areas, which gradually increase in size during storage. The problem is apparently related to localised oxidation at the sausage surface. The affected areas are characterised by a low sulphur dioxide content and by fat with a high peroxide content. The condition has also been associated with the use of pre-soaked rusk, with an implication of increased bacterial activity on the rusk particles. White spot can be effectively prevented by the addition of the proprietary reducing agent Ronoxan D20® at levels of 0.5–1.0%.
- **Blue spot**. This is an optical effect sometimes caused by pieces of cooked pork rind beneath the sausage skin. Iodine-containing sanitisers have also been known to give blue colours by reaction with the starch present in rusk.
- **Artificial colours**. Red specks may result if colour is not properly dispersed. Rinds may be stained preferentially by Red 2G. Addition of too much artificial colour may result in an undesirable colour in the cooked

product. Dye from identification stamps as used by meat inspectors, for example, may be encountered occasionally.
- **Colour dilution effect**. Additives such as soya may reduce the colour intensity and result in a pale product. Fat has a similar effect, especially if finely divided. 'Creaming' of the fat in a sausage mixture may occur as the sausages pass through high speed filling machines, resulting in a pale surface appearance (page 139).

Emulsion sausages

Frankfurters, etc.

The most common form of emulsion sausage in the UK is the *frankfurter*, made in the following way:

- Meat is very finely chopped. Much of the fat is therefore in the free form; it is bound in stable form by the strong lean meat–water–salt mixture produced simultaneously by the fine chopping.
- The mixture is cured with sodium nitrite to give a pink colour; this is not essential for other emulsion sausages.
- The mixture is filled into disposable casings, usually cellulose, cooked to heat-set the meat, and the casing is then removed.
- When fully heat-set on cooking, a good frankfurter will break with a 'snap'.

Since it is fully cooked the frankfurter should have a low microbial count. However, it is still intended for more or less immediate consumption, unless canned or frozen.

Other emulsion sausages include:

- wiener (Vienna sausage) – very similar to frankfurter
- German Bruhwurst (generic term = 'scalded sausage') – Frankfurter, Bierschinken, Jagdwurst
- French saucisson – includes saucisse de Strasbourg.

Brat production

The finely chopped mixture of lean meat–water–salt and fat is called in English a 'brat' (although the original German word refers not to chopping but to roasting or grilling – as in 'Bratwurst'). Many continental speciality sausages consist of a 'brat' with pieces of chopped meat or fat, spices etc., dispersed through it.

Formulation of 'Brat' (approximate):

lean meat	35%	water	28%
fat	35%	salt	2%

Notes

(a) Most countries impose a legal limit of 50% fat in the meat portion. If other meats, fats, etc., are to be included in the complete sausage, the 'Brat' formulation may be adjusted accordingly. Rind may be substituted for lean pork meat, e.g. 10% is permitted in Germany. (See also above, pages 133, 136.)
(b) A high water content is essential for making the 'Brat'. The UK standard for canned frankfurters requires 70% meat content; the remainder is water and salt.
(c) Salt content is normally limited by flavour. Phosphate may be included (but is prohibited in Germany). Other salts (sodium citrate, lactate) may be added to improve water binding without increasing the salt taste (page 3).

Chopping procedure (German recommendations)

- *Either* Start with lean meat + salt (preferably pre-salt the meat: chop coarsely and leave for 1–2 days at 5°C (41°F) or below). Chop, add water gradually. Add fat and chop at high speed until fine. (Note the similarity of this to the process described on page 133).
- *Or* put all ingredients into the chopper and chop at very high speed until fine. This process is suitable for large scale production. Similar results may be obtained in other equipment, e.g. a KS mill.

Keep knives sharp and very clean, to cut cleanly and reduce friction. Final temperature after chopping should not be too high, 22°C (72°F) the absolute maximum, but some consider 18°C (64°F) or lower should be the limit in a very high speed chopper or mill.

Dried and fermented sausages

See Chapter 9.

Spreadable meat products

The main features of these products are:

(a) The meat is cooked and comminuted during the manufacturing process. Pre-cooked meat (e.g. canned meat) may be used as an ingredient. The product may also be heated after the mixing and cooking stages (e.g. meat pastes are filled into jars, etc., then sterilised). Because of the comminution after cooking, the product is not held together by lean meat binding of the kind which occurs in sausages etc.
(b) Product integrity and spreadability depend on the presence of fat and

weak gel, holding the cooked meat particles together. The fat or gel breaks down easily under the forces which cause spreading; the fat also acts as a lubricant to help meat particles slide over one another.

Liver sausage

Recipes may contain rather variable amounts of liver and lean or semi-lean meat, with some ingredients high in connective tissue (e.g. pork jowls, rind, lungs), which provide gelatin in the finished product.

Typical recipes

Meats etc:	Liver	50	50	20
	Meat (80% VL)	10	25	15
	Meat (50% VL)	40		
	Added fat		15	50
	Rind or lung		10	15
Other ingredients:	Salt	1–2%	of the total recipe	
	Water, ice or stock	0–40%	of the total recipe	
	Rusk	0–12%	of the total recipe	
	Onion, spices, etc.	to taste	of the total recipe	

The liver and meat ingredients are lightly pre-cooked, finely minced or chopped with the other ingredients, filled into casings, cooked in water or moist air at 74–85°C, (165–185°F) and cooled.

Sliceable liver sausage may be made in a similar way but without pre-cooking the lean or semi-lean meat. There is therefore some development of meat binding during cooking to provide rigidity and sliceability.

UK regulations require 50% minimum meat content in liver sausage.

Pâté, terrine

Pâté is the French term for a spreadable meat paste. It is usually understood to apply to a product with high meat content; UK regulations permit the term to apply to any spreadable meat product with 70% minimum meat content. 'Terrine' is, strictly, an earthenware vessel in which the pâté may be cooked.

Typical recipe

Liver	26
Pork belly	26
Back fat	22
Pork rind	8

Onions	5
Garlic powder	0.3
Salt and phosphate	2.3
Water	10.4
Sodium nitrite	200 ppm (input)

The rinds are cooked and the meat is lightly cooked, then all the ingredients are finely chopped together in a bowl chopper, filled into bowls (terrines) or moulds and baked for $1\frac{1}{2}$–3 hours depending on size and the shelf-life required (page 146).

Meat paste, meat spread

These products are similar to pâtés and of intermediate texture. They are commonly made in the UK with meat content close to the regulation minimum of 70%. The non-meat portion usually consists of rusk and water ($1+3-3\frac{1}{2}$), or other suitable filler such as soya concentrate or sodium caseinate. The product is commonly heat sterilised after filling into jars or cans, e.g. for 3 oz (75 g) glass jars: 60 min at 115°C (240°F), $F_0 = 10$.

Potted meat

This is minced, cooked meat, filled into pots or jars, etc.; it is not normally finely comminuted and is therefore only coarsely spreadable.

UK regulations require 95% meat content, unless the product is labelled to indicate ingredients other than meat.

Potted hough and **potted head** are Scottish products of lower meat content. UK regulations require no declaration of meat content.

Typical recipe

Beef hough (shin beef or head meat)	50
Pork rind	15
Stock (from cooking the above)	28
Salt and seasoning	7

The meat and rind are minced after cooking, mixed with stock, etc., and placed in containers.

Problems in spreadable products

Water separation

To avoid this:

- cool to below 21°C (70°F) before filling
- re-mix before filling

- ensure sufficient rusk in the recipe to absorb all the free water from the first cooking (e.g. 1 part rusk for 3½ parts water).

Fat separation

Fat in these products appears to be held by adsorption or entrapment between cooked meat particles. There is no meat matrix of the kind described on page 000 and there is no significant emulsification.

Above about 15% fat content in the product, fat separation occurs on cooking. A level of 17% may be a practical maximum if a small amount of visible separation can be accepted. In some products, e.g. pâtés in terrines, a significant fat layer on top may be acceptable or even preferred. Here the fat layer can contribute to increased shelf-life by sealing the rest of the product from contamination. Attempts to incorporate more fat without separation, e.g. by the use of fatty tissue, have not been successful.

Shelf-life

Pâté and similar products
If the product is exposed to the air, shelf-life is usually terminated by the development of stale flavour and aroma and a grey colour. These are due to fat rancidity and oxidation of uncooked meat pigments.

Vacuum packing, canning, etc., therefore extend shelf-life. Frozen pâté has a short shelf-life – see page 000 for the acceleration of rancidity by freezing.

Nitrite extends the shelf-life of pâté; e.g. at 5°C (41°F):

No nitrite	0–1 week
200 ppm input nitrite, giving approximately 50 ppm residual nitrite after manufacture	3–4 weeks

Polyphosphate (0.3%) extends shelf-life slightly but makes the pâté softer.

Meat pastes
Shelf-life is often determined by darkening of the surface meat in the headspace in the jar, can, etc. This darkening is accelerated by iron. Where the closure of a jar is the source of iron, shelf-life may be extended by the inclusion of an iron-free paper disc.

Effects of heat processing

Sterilising
Meat pastes and spreads, packed in jars or cans, are sterilised to $F_0 = 10$ or more. Shelf-life is therefore long, usually terminated by chemical changes as noted above.

Pasteurising
Open pack pâtés, terrines etc., are pasteurised. For a terrine of 2.5 kg or more allow $2\frac{1}{4}$ h at 148°C (300°F).

Microbial counts diminish with increased cooking time, but organoleptic shelf-life is not affected.

Artificial colours

There are now no colours with satisfactory performance in sterilised meat pastes, still permitted in the EU.

Luncheon meats, meat loaf, etc.

See page 165.

9 Cured Meats

PROCESSES

Injection

All curing processes of non-comminuted meat, except for dry curing, involve as a first step the injection of a brine solution into the meat. The brine contains the necessary curing agents salt and nitrite, usually also nitrate; other ingredients may be included as required by particular recipes.

There are three main methods of injection: multi-needle injection machines, hand 'stitching', and arterial injection.

Multi-needle injection machines

This is now the most commonly used injection method, especially for pieces of boneless meat. Models are available for use with bone-in sides or cuts.

Hand 'stitching' or 'pumping'

This was the traditional method before the invention of multi-needle injectors, and is still used in some factories. Single-needle injectors, the best ones fitted with brine meters and pressure gauges, are used.

For bacon in the UK, operators follow a strict pattern of injection points into the bacon side (see the BMMA Standard for the Production of Bacon and Bacon Joints, 1996) to ensure optimum distribution of the cure. Pressures of 70–90 psi (5–6 bar) are used.

Arterial injection

First introduced for hams in the USA in the 1930s, this method is still used for some high quality hams and for tongues. Brine is injected into the open ends of arteries and is carried through the meat by the arteries and capillary system. The following conditions must be observed:

- The arteries must be undamaged by previous processing and it must be possible to find them; skill is therefore required.
- The animal must be properly bled so that the arteries are more or less empty of blood.

- Pumping pressures must be carefully controlled: 30 psi (2 bar) minimum for effective distribution, 40 psi (2.3 bar) maximum to avoid bursting the arteries.

Tanking or immersion curing

The Wiltshire bacon process and some of its many modern derivatives are described on pages 154ff.

Dry curing

Dry salt or salt mixed with saltpetre (sodium or potassium nitrate) is incorporated into the meat, often with gentle rubbing by hand, daily over several weeks. Up to 3% salt overall is taken up, the concentration being naturally much higher near the surface. Between treatments the meat is hung up in a cool dry atmosphere and moisture is slowly lost from it; the final weight loss may be 10–25%, depending on the conditions.

The process should be commenced as soon as possible after the initial chilling of the carcass, before any significant build-up of spoilage bacteria. Good hygiene remains necessary in the early stages of curing when the moisture content of the meat is high and the salt content low; preservation then is assisted by the solid salt surrounding the product. Warm conditions assist the penetration of curing salts but are less good for hygiene; a suitable compromise temperature is 5–7°C (41–45°F).

Cutting should be minimised to avoid contamination of the interior of the meat. If the product is required boneless, de-boning should be delayed until after curing is complete.

Massaging, tumbling

Penetration and equilibration, especially in relatively small pieces of meat, are greatly speeded by the mechanical action. Myosin extraction at the surfaces also takes place, with binding into cohesive sliceable pieces on heating. The combination of massaging or tumbling, followed by heat treatment of products filled into moulds or casings, is therefore used for the production of re-formed cured-meat products.

Massaging means the gentle rubbing of one meat surface on another, or on a smooth surface such as the massager wall, without loss of contact. Typical massage machines are tanks containing slowly rotating, smooth-shaped paddles (Fig. 9.1). Tumbling is a more vigorous action in which the meat is struck by paddles, baffles, etc., or allowed to fall from the upper part of a rotating drum (Fig. 9.2). The earliest machines included butter churns, cement mixers, etc.; many designs of rotating drum are now available.

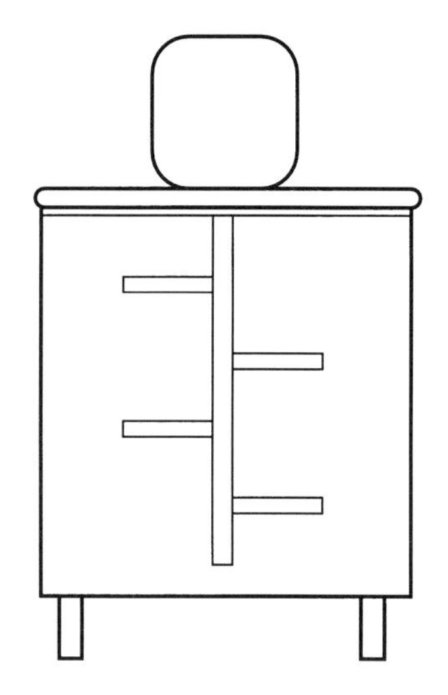

Fig. 9.1 Massager.

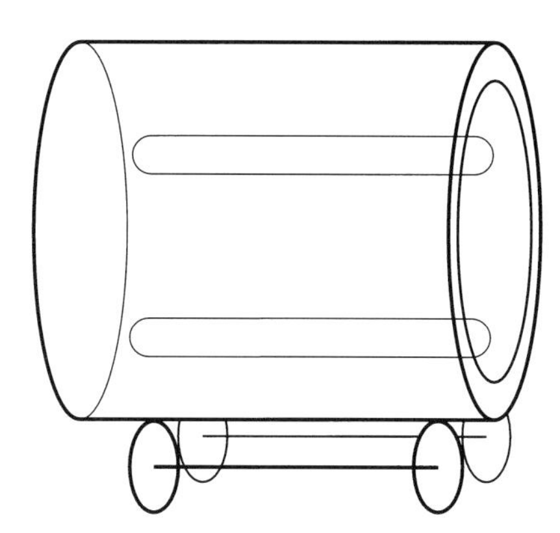

Fig. 9.2 Tumbler.

Massaging or tumbling usually follows multi-needle injection. Wide variations in operating conditions are possible; two main variants are common:

- Massage or tumble for a period (e.g., $\frac{1}{2}$–1 h) rest 16–24 hours; massage or tumble for another period.
- Massage or tumble for short periods (5–10 min) once each hour for 8–24 hours.

Results are judged in terms of:

(a) yield of cured product, before and after cooling;
(b) adhesion of pieces of meat together on cooking, to give an intact, sliceable product;
(c) loss of fibrous structure in the meat.

Too little mechanical action reduces (a) and (b); too much increases (c) and may reduce (b).

Smoking

Smoke from hardwoods (oak, beech, hickory) has traditionally been considered best but in practice any non-resinous wood may be used.

Present-day modifications to the old method of hanging meat in the domestic chimney include the following:

- Smoke generation from shavings or sawdust in special generators, usually attached to cooking cabinets; the smoke is more intense if the wood is wetted.
- Where smoke generation is in a separate unit from the cooking chamber which can be controlled independently of temperature and humidity, the meat may be dried and/or cooked, as required, before smoking in the same equipment.
- Smoke temperatures may be adjusted for 'cold smoking' (typically around 35–50°C, 95–122°F) or 'hot smoking' (typically over about 80°C, 180°F).
- Harmful substances (e.g. benzpyrene which is carcinogenic) can be removed from the smoke by including water sprays or electrostatic precipitation between smoke generation and deposition on to product.
- Smoke concentrates or essences may be used instead of natural smoke; they are sprayed on, sometimes assisted by electrostatic precipitation from a colloidal spray.

Smoking may be from 4–12 hours, depending on the conditions and product required. The internal temperature of the meat may rise to around 35°C (95°F) during hot smoking. This should be reduced as rapidly as possible afterwards.

PRODUCTS

Bacon and ham:

Bacon

Bacon is cured pork. In the UK and generally in Europe it may be from any part of the pig. In North America 'bacon' usually refers specifically to belly bacon (see below). Bacon may be smoked or unsmoked ('green').

Wiltshire side

Wiltshire side is the half carcass (without head) cured in one piece. It is subsequently cut into joints and boned out for retail sale.

If the leg (8 and 9 in Fig. 9.3) is cured after being cut from the side, it is not called gammon but ham. Any of the other pieces may also be cut from the carcass and boned before curing. Such pieces may be called **block bacon** or **block-cured bacon**; the cured pieces have similar names to the above.

Ham

A ham is the upper leg and buttock of a pig, uncured or cured. Ham is the cured product made from this. In the UK it is usually, but not necessarily, cooked.

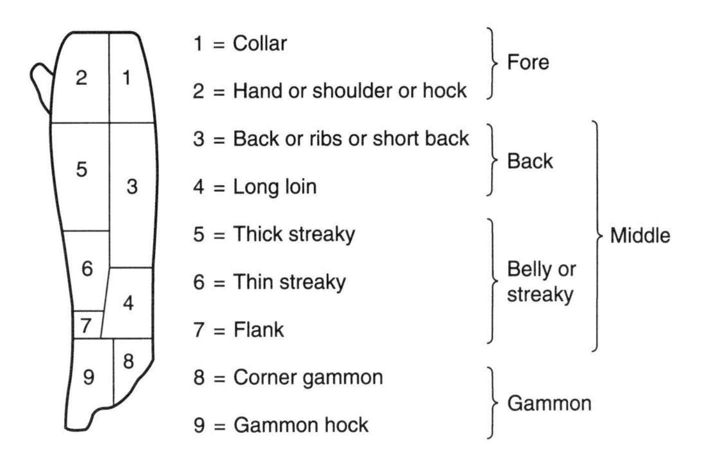

Fig. 9.3 Bacon cuts

Gammon or gammon ham

This is the ham cut from a Wiltshire-cured side. In the UK this also is usually, but not necessarily, cooked.

Raw ham

Raw ham includes York ham, Westphalia ham, jambon de Paris, prosciutto (Italian) and others. It is usually dry-cured (see below) but not cooked.

Re-formed ham, re-structured ham

These are names for pieces of cured ham meat reassembled into rolls or blocks. Similar products made from meat from other parts of the pig may not be called 'ham' or any other name including that word; 'cured pork shoulder' and other terms are acceptable.

The Wiltshire process is a traditional method of bacon manufacture in which:

- The only ingredients are a salt brine containing the curing salts potassium (or sodium) nitrate and nitrite; in some modern versions the nitrate salt is omitted.
- Curing progresses in three stages: brine injection, immersion in a 'cover brine' of brine, and a maturation period.

Speck

Speck is the name used in continental Europe for cured, heavily smoked, pork back fat or fatty belly.

Wiltshire process

The process is applied to whole sides or to cuts of pork (middle, back, streak or fore-end). The curing stages are:

- **Injection** of brine. Traditionally this was done by hand 'stitching' or 'pumping' following a regular pattern to ensure optimum distribution. It is now almost always done by multi-needle injection machines. It is common also to pack extra salt into the 'shoulder pocket' where the blade bone has been removed (to prevent spoilage in this region before curing is completed, due to increased bacterial contamination during the de-boning process).
- **Immersion** of the injected meat in tanks of cold brine (0–5°C, 32–41°F, previously for about 5 days, now usually 3–4 days). Further salt may be sprinkled on the surfaces of the meat as it is stacked in the tanks.
- **Maturing** of the sides out of brine, stacked in a cellar (0–5°C, 32–41°F, previously up to 14 days, now usually 3–5 days).

(Full details are given in the BMMA (1996) Standard for the Production of Bacon and Bacon Joints.)

Brines

Injection brines
'Live' brine from a curing tank may be used (see below) but it is more usual to make up fresh as required, with similar composition to the immersion brine or, commonly, slightly diluted.

Wiltshire brines
The immersion brine is a 'live' brine, traditionally made with saturated salt plus sodium or potassium nitrate and containing a large population of active bacteria which:

- grow readily under the conditions of high salt concentration, presence of nitrate and low temperature;
- produce nitrite continuously from the nitrate in the brine;
- suppress the growth of any other bacteria which might cause spoilage.

Sodium (or potassium) nitrite is now almost always included to supplement that produced microbiologically.

A typical Wiltshire brine contains:

Salt	24–25% w/w (saturation = 26% w/w at 0°C, 32°F)
Sodium nitrate	0.4% (4000 ppm) (or potassium nitrate 0.5%)
Sodium nitrite	0.1% (1000 ppm)
Soluble solids	(from meat previously cured in the brine)

Micrococci and lactobacilli	(Total count by microscopy 10^7–10^8 per ml; 5-day count at 22°C on 4% salt agar below 5×10^5 per ml)
pH	6.0–6.5

Such a brine is microbiologically stable provided:

(a) The salt content is maintained close to saturation; this requires the regular addition of salt to replace that which is absorbed by the meat. Failure to ensure this may permit unwanted bacteria to become established which may decompose nitrite or otherwise interfere with the balance of the system. The salt concentration should be monitored at least daily using a hydrometer (salometer) and more salt added when required.
(b) Nitrite may be added as necessary to ensure the correct concentration.
(c) Nitrate is added regularly to replace that consumed in forming nitrite.
(d) The brine is kept cold; only meat which is already chilled should be put into it.
(v) The brine is kept aerated; in normal operations the immersion and removal of meat will ensure sufficient aeration but if a brine is left unused for more than a day or two it should be pumped, stirred or splashed occasionally.

Regular chemical analysis (e.g. at least weekly) is necessary to control (b) and (c). During use, curing salts are withdrawn from the brine into the meat, while water from the meat dilutes the brine. Continuous adjustment is therefore necessary. Regular microbiological analysis is also helpful.

It is an objective of the Wiltshire process that an immersion brine, carefully managed in this way, should remain continuously active and stable and never need to be removed or destroyed. New brine, if ever required, may be made with a proportion of a satisfactory old brine, to provide the correct microbial flora and some dissolved meat solids.

Maturing

In the Wiltshire process this stage is necessary:

- to drain off excess liquor, and
- to allow curing salts to equilibrate through the meat.

Careful stacking of the sides is important to allow drainage without distortion. The meat should be kept cold (5°C, 41°F, or below); therefore traditionally the maturing stage was done in cellars. Relative humidity must be low: about 85% RH is satisfactory and is achieved in practice by keeping the floor wet with brine, not water, at all times. (Saturated salt solution is in equilibrium with 75% RH.)

Modifications, simplified and rapid curing processes

'Fresh' brines

Instead of using immersion brines continuously maintained at the appropriate chemical and microbiological composition, sometimes for very long periods, it may be preferable to make up fresh brine whenever required. With no need to maintain near saturation of salt to ensure the correct bacterial population, the salt contents of fresh brines may be lower than in a Wiltshire brine. This permits somewhat faster penetration into the meat, therefore shorter immersion times; the period may be reduced to 1–2 days. There is also an increased uptake of water, which above certain limits may need to be declared when the product is sold: see page 200.

Typical concentrations are:

Salt	18–22%
Sodium nitrite	0.06–0.1%
Sodium nitrate	0.15–0.2%

Satisfactory bacon can also be made with brines containing nitrite but no nitrate, since there is of course no need for the microbial conversion of nitrate to nitrite.

As with the Wiltshire process, it is very important to keep close control of the chemical composition of the brines and of the finished product, with regular analyses of the salt and nitrite contents. Production records should also be kept of the ingoing and outgoing weights, and therefore the yield, of every batch. This is to confirm the chemical analyses of composition and to show the amount of water absorbed or lost – which in the UK at least will be required to justify declarations on the label of the final product (see page 200).

Brine injection without immersion

The purpose of the immersion or 'tanking' stage in the Wiltshire process is to allow more uniform distribution of the curing salts to be attained throughout the meat. Any measures which will allow uniformity of composition without the need for an immersion stage and the associated problems of control of brine composition, the labour of handling and the time requirement, may simplify the whole process and reduce its cost. Such measures include block cures, bag curing and slice curing.

Block cures

This term is usually taken to cover bacon made from pre-cut pieces of pork by injection processes, usually with a multi-needle injector. The injection process may be, and usually is, followed by a maturing or equilibration stage (usually 1–3 days) before smoking, slicing, cooking, etc.

The equilibration stage may be carried out with the meat in an immersion brine or 'cover brine' of similar composition to the injection brine.

All brines are freshly made.

'Bag curing'

After injection and draining (e.g. for 24 hours) the meat is packed in impermeable plastic bags in which equilibration and curing are allowed to continue until the product is removed for slicing, packing or other processing.

Slice curing

In this case, the meat is pre-sliced, then the slices are immersed in a bath of curing brine for a short period (less than 1 minute) to pick up the required amounts of curing salts. The process can be automated and can give close control of composition. It was operated for some years by Unilever under patent (Brit. Pat. No. 848014) but was later discontinued because, it is said, of difficulties in packing the cured slices as neatly as if they had been sliced after curing.

Other ingredients in curing mixtures

These may include:

- phosphates – to improve water-holding, especially in massaged products;
- sugar – for flavour; reducing sugars such as dextrose or lactose also have a beneficial effect on the colour of cured meat;
- milk powders – for flavour, also a source of lactose;
- monosodium glutamate (MSG) – flavour enhancer;
- hydrolysed vegetable protein – flavour;
- smoke or smoke essences – flavour, see below;
- ascorbate or erythorbate – curing aid;
- soya or caseinate – to improve water-holding and texture in products of moderate or low meat content.

Packaging of bacon

Slicing

The accuracy of slicing is affected by several factors, as follows.

Temperature control

With hand-operated or small-capacity automatic or semi-automatic slicers, there are few special problems. With high-speed automatic slicers, the evenness of slicing depends greatly on the rigidity of the blocks being sliced, which depends on temperature. Before slicing, the bacon should be lowered to a temperature as close as possible to −2°C (28°F). At lower temperatures ice will form and make slicing more difficult. At higher temperatures the meat is softer.

Fat content and fat composition

Softness of the fat is the major factor in causing the meat to be softer or less rigid. See page 18 for background information. The softness of the body fat of pigs may be very variable and is influenced by the diet fed to the live animals. Difficulties in slicing may be traceable to this cause; medium- to long-term corrective action is possible at the farm level.

Shaping of blocks

Better performance is obtained from high-speed slicers, and better portion control of the product is possible if the meat is shaped to uniform cross-section before slicing. This can be achieved with various meat or bacon presses, which compress blocks of meat (usually at about −2°C, 28°F) within regularly shaped chambers.

In addition to the usual dangers in meat products handling, there are possibilities of microbial contamination of sliced product due to:

- build-up of organisms on slicers, tables, conveyors, etc., if there are any defects in cleaning procedure;
- possible cross-contamination between cooked and uncooked products.

These possibilities must be controlled.

Packaging

Whole Wiltshire sides were traditionally packed in muslin stockinet. Nowadays bacon is mainly prepared by the manufacturer as primal cuts, boned and vacuum packed, and may be pre-packed centrally into vacuum packs for further distribution before sale. Some bacon is sliced from the primal cuts in supermarkets, etc. and pre-packed into loose bags or packages without vacuum, but the majority is vacuum packed.

The extra shelf-life to be expected from vacuum packaging may be offset at the point of production by a reduction of salt content to provide a milder-tasting product.

See page 84, 85 for the conditions of vacuum- and gas-packing.

Pasteurising

Some bacon is part cooked, hot smoked, etc., which improves shelf-life. For practical smoking conditions see page 151.

Hot smoking produces a small pasteurising effect; additional heating may also be provided. The bacon is heated, with or without smoke, in the block form before slicing, e.g. on racks in air-heated cooking cabinets. Heating in a finished vacuum pack might be effective microbiologically but produces unsightly packs due to cooking losses; the slices may also stick together. The total amount of heat necessary for effective pasteurising has not been studied in this context.

Frozen bacon

This is very prone to rancidity unless packed under perfect vacuum. Frozen storage life is probably not more than six months, and may be less (see page 94).

Ham production

The principles involved, and some of the methods used, are similar to those for bacon, with certain additions described below.

Cooking of ham

Cooking alters the palatability of the product. It may also increase the shelf-life, because of its pasteurising or sterilising effect.

Products with high salt content – say 5% salt-on-water or higher – are shelf-stable at room temperature, whether cooked or not. Products with lower salt content, e.g. most injection-cured products, are not shelf-stable for more than a few days unless they receive some heat treatment.

There is conflict between two sets of requirements:

- to give sufficient heat to pasteurise or sterilise the micro-organisms present;
- not to give too much heat, so as to avoid large cooking losses or too much softening of texture.

Even when nitrite is present, the heating necessary for the first requirement may be too great for the second requirement. The problem is further complicated in the case of large pieces of meat such as whole hams or cured meat in large cans, because of temperature differences during cooking between the outside and centre of the product. In practice the following solutions have emerged:

- Small cans (up to *c.* 1 lb, 450 g), shelf stable at room temperature, are given heat processes similar to those for luncheon meats ($F_0 = 0.1–0.7$ approximately (page 000).
- Large cans, plastic packs, etc., $1\frac{1}{2}$–6 lb (600 g–2.5 kg) or more, including whole hams, pear-shaped cans of *c.* 6 lb capacity and long square-sectioned 'Pullman' cans, are pasteurised only and the product must be kept under refrigeration (0–5°C, 32–41°F). Hams up to 15 lb (6.8 kg), cooked in metal moulds, may also be included, although this is not strictly a canning type of process since post-process contamination cannot be avoided.

Heat process is measured by the centre temperature attained. This is not normally less than 66°C (150°F). The meat is likely to be undercooked at 65°C (148°F).

Table 9.1 shows some typical figures for cooking loss in hams (9–14 lb approx.) cooked in moulds to various centre temperatures, with comments made at the time on the degree of cooking.

Table 9.1 Cooking temperatures of ham (data from Ranken, 1984)

Centre temperature °C	°F	Cooking loss %	Remarks
61	141.5	2.6	Half raw
64	148	4.4	Slightly undercooked
66	150.5	9.6	Slightly undercooked
69	156.5	7.8	OK
69	156.5	11.1	Undercooked
70	158.5	9.7	Slightly overcooked
71	160	12.1	OK

The cooking loss is affected by other factors, including processing temperature and pH. The effect of the latter is quite complex and not reliably predictable in individual cases. In all these products it is common practice to add up to 2% dry gelatin before cooking and disperse it well, to ensure that any cooked-out liquor forms a jelly which solidifies on cooling.

Cooling of cooked hams

This stage must be carried out thoroughly and as quickly as possible, especially with large hams or other pieces, so that excessive microbial growth is avoided at the centre of the meat which remains warm for longest. For fully cured meats (2.5% salt-on-water, 100 ppm input nitrite) the centre temperature should be brought down to 5°C (41°F) in not more than 10 hours.

Slicing and packing of ham

Hygienic requirements are similar to those for bacon (pages 158, 159), with the addition that cross-contamination from uncooked meats *must* be eliminated (page 000). High-speed slicing is less common than with bacon and the technical problems of accurate slicing may be less troublesome (page 157).

Composition of the finished products

Bacon

At pick-up levels of up to 10%, especially if no immersion brine is used, average product composition can be calculated reasonably accurately from brine composition and the proportion of brine taken up. At higher levels of pick-up the calculation is less accurate because of a tendency for any brine lost from the meat to have a different composition from the injected brine. In practical manufacture it may be difficult to measure pick-up reliably and considerable variation may be found. Practical experience suggests that composition becomes more uniform if an equilibration stage is allowed; at any rate, shelf-life of the product is considered to be more reliable.

In a typical case, a 10% pick-up of brine, of the composition on page 156, gives the following composition in the bacon:

	Theoretical	Practical
Salt	2.5%	3%
Sodium nitrate	500 ppm	250 ppm
(expressed as sodium nitrite)	400 ppm	200 ppm
Sodium nitrite	200 ppm	80–150 ppm

The 'practical' salt content is high because:

- additional salt is absorbed from the brine for reasons which are not understood
- dry salt is added to the meat after injection.

In bacon made with a 'live' brine the nitrate and nitrite contents may be different from the above, depending on the state of the microbial nitrate reduction processes in the brine at the time of use.

There is considerable variability about the average, within the bacon (see page 165). There is no evidence that Wiltshire bacon is less variable than other kinds.

The microbial flora from a Wiltshire brine remain active on the finished bacon and may continue to convert the residual nitrate into nitrite during storage. Thus, although a bacon might be made satisfactorily, it could nevertheless exceed the legal maximum nitrite content after it has been made. However, since the nitrite content itself diminishes on storage (page 53), the effect is temporary and not usually a problem.

Fate of added nitrate in block-cured bacon

Nitrate may be added to block cures in the hope that, as with Wiltshire bacon, it may be reduced to nitrite during storage of the product, thus prolonging the shelf-life. Reduction to nitrite will occur only if the necessary micro-organisms are present in the environment where the bacon is made; if Wiltshire bacon is or was made in the same factory, the correct microbial flora may be present but in other premises this cannot be guaranteed. It has been found that when nitrate-containing bacon was made in factories where bacon had not been made before, either the nitrate present was not reduced and its presence had negligible effect on shelf-life, or the nitrate in the bacon encouraged spoilage bacteria (Enterobacteriaceae) and shelf-life was therefore shortened.

Nitrate is probably best avoided in new products made by block curing.

Dry cured hams

Some typical figures for a York ham (bone-out) are:

Moisture	31.0%
Protein	5.0%
Fat	49.0%
Salt	2.9%
Apparent meat content	118.6%

The fat content may be expected to be variable. The salt content may be rather unevenly distributed.

Other hams. These will have lower apparent meat contents, depending on the amount of brine taken up and retained in curing and cooking.

Injection-cured hams. The expected average content of salt, nitrite, etc., can be calculated, as with bacon, from the brine composition and the pick-up proportion where this is known.

The combination of injection, tumbling or massage and careful control of cooking can produce products with high proportions of brine retained after cooking. In the past this led to legal difficulties when products with low water content (e.g. dry-cured ham, which may lose water during production) or with high added water contents might all be called by the same name: 'ham'. This is likely to be resolved under current UK legislation which requires the declaration of any additional water over that normally added in the curing process.

For calculation of meat content or added water content from an analysis, see pages 197–200.

Variability of composition

Even when made under careful conditions there may be wide variations in composition within batches or units of a cured meat product.

Example 1
Three canned hams from each of two manufacturers were divided systematically into 24 portions; means and ranges of the analytical values found for the 24 portions of each ham are shown in Table 9.2.

Example 2
Seventy-six ½-pound samples of bacon from the same factory were selected at random. Figure 9.4 shows the relationship between salt and nitrite contents of these samples.

Tongues

The normal process for curing tongues is as follows. Trim off excess fat and glands, and remove hyoid bone if this can be done easily. Pump with curing brine. Ox tongues may be artery pumped (35 psi, 2.4 bar, appears to be the optimum pressure). Smaller tongues are usually pumped by multi-needle injection but this may give decreased overall yield, even when phosphates are used. Cure in a cover brine for approximately 2 days at 0–2°C (32–41°F). Drain. Cook until the skin can be removed. The tongues are normally simmered at 97–99°C for *c.* 2½ hours (ox tongues) or 3½ hours (lamb tongues). Remove skin, and the hyoid bone if still present. Pack into cans (gelatin may be added). Heat-process to $F_0 = 0.1–0.7$.

Yields can vary widely, for reasons not yet clearly established. The use of polyphosphates may be only slightly beneficial.

The cured colour may be uneven, again for reasons not well understood.

Problems encountered are:

- yield variations (not well understood)
- uneven cured colour.

(The use of tongues is not prohibited under UK Regulations relating to BSE.)

Table 9.2 Variability in composition of canned ham (data from Ranken, 1984)

	Manufacturer A			Manufacturer B		
Ham number	1	2	3	4	5	6
Time of storage (months)	0	1	4	0	1	4
Sodium nitrite (ppm)	34 (8–70)	12 (7–28)	7 (2–12)	90 (65–116)	140 (30–47)	29 (22–42)
Salt (%)	4.18 (3.2–4.9)	4.07 (3.6–4.6)	3.22 (2.9–3.3)	4.44 (3.4–5.7)	3.54 (2.8–4.5)	3.36 (3.1–3.7)
Water (%)	70.03 (67.2–72.1)	69.03 (62.8–71.5)	71.78 (68.1–75.2)	68.40 (62.4–72.9)	67.59 (58.0–71.7)	67.60 (60.4–71.1)
Salt in water (%)	5.96 (4.8–7.1)	5.91 (5.5–6.6)	4.48 (3.9–4.7)	6.49 (5.4–8.0)	5.24 (4.5–6.6)	4.96 (4.7–5.2)
pH value	6.36 (6.2–6.8)	6.12 (6.0–6.2)	6.42 (6.3–6.6)	6.34 (6.2–6.5)	6.28 (6.1–6.4)	6.35 (6.3–6.4)

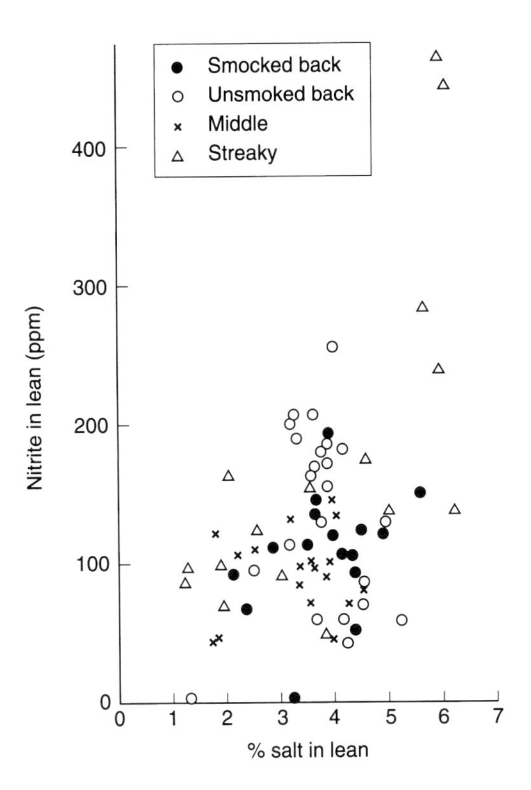

Fig. 9.4 Variability in composition of bacon (data from Ranken, 1984).

Typical compositional data for canned and frozen ox tongues

	Canned ox tongues (8 oz cans) 18 samples	*Frozen ox tongues* 107 samples
pH	6.16–6.72 mean 6.36	5.7–6.5 mean 6.05
Water %	58.5–72.8 mean 67.4	
Salt %	2.03–3.57 mean 2.76	

Nitrate (as $NaNO_3$) at 93–269 ppm was also found in samples of tongues cured in brines to which nitrate had not been added (see page 56 for nitrate contents of other uncured meats).

Luncheon meats

Luncheon meats are finely chopped mixtures which may have high contents of fat and connective tissue. They are normally cured and canned. Meat loaf usually has a lower meat content, may be more or less coarsely chopped and is not usually cured.

To a large extent these products are designed to use up trimmings, etc. from the other processes in a factory. The recipe and process should therefore be simple and adaptable to fairly wide variation in raw materials.

Regulations regarding meat content in the UK are:

- Meat content must be 65% if described as 'meat loaf' or 'meat roll', 80% if described as 'meat with cereal' or equivalent; it may be higher.
- Not less than 60% of the meat content must be lean meat.
- Rind (as rind less trimmable fat) may be substituted for lean at 10% of the total pork meat content (pages 133, 136).

Technology

Chopping

This follows the principles discussed earlier for making stable sausages (page 47). If cooked meat is included in the recipe, its poor binding properties must be allowed for. Farina (potato starch) may be used as a water-binding agent: it is particularly effective in this type of product.

Curing

The use of nitrite gives a pink colour to the finished product and also permits the use of a low heat process (see below). At an input level of 75 ppm, sodium nitrite is sufficient for both purposes provided it is well mixed in during the chopping process. If cured meat is used in the recipe, its nitrite content should be allowed for; its effect in the final product is the same as that of nitrite added with uncured meat. Below 10 ppm input nitrite, colour development is incomplete. At intermediate nitrite levels the use of ascorbate is helpful (page 74). Erythrosine has been used as an artificial colouring agent but only Allura Red is now permitted in the EU.

Heat process

Given meat of good microbial quality (clostridial spores not more than 1/g) the recommended minimum heat process is:

Minimum content of sodium nitrite	75 ppm input
Minimum salt content	3.5% on moisture
F_0	0.1

There is evidence that a heat process of the same F_0 value is more effective if the processing temperature in the retort is higher (e.g. processing at 115°C (240°F) is better than processing to the same F_0 value at 110°C (230°F)). Under these conditions in a well-made product, in-can cooking losses should be small or negligible.

Dried and fermented sausages

These are not commonly made in the UK. European varieties include:

French saucisson sec e.g. rosette, Jésus
German Dauerwurst, Rohwurst, e.g. Zervelat, Braunschweigerwurst
Italian crudi stagionati, e.g. salami, coppa

Production is as follows:

(1) The mixture (just frozen) is coarsely chopped so as not to encourage strong water binding by lean meat, because of the drying process which is to follow later. Salt is used but no phosphate. Fat must be cut cleanly without smearing. (Chop frozen – since the product will not be heated to 40°C the problem of fat loss does not arise).
(2) A bacterial ferment (starter culture) is added to initiate two fermentations simultaneously:

(a) nitrate $\xrightarrow{\text{micrococci}}$ nitrite

(b) carbohydrate (sugars, etc.) $\xrightarrow{\text{lactobacilli}}$ lactic acid

The fermentations are established under conditions of high temperature and high humidity, e.g. 26°C (79°F), 95% RH.
(3) Under these conditions fermentation (a) occurs rapidly at the beginning of the process; the nitrite formed prevents spoilage while there is not yet sufficient lactic acid from fermentation (b) to suppress spoilage organisms. The fermentations cease after 3–7 days, when enough lactic acid has been formed, with pH about 4.5, to inhibit further growth of lactobacilli, or when the carbohydrate supply is exhausted.
(4) Humidity and temperature are then reduced, e.g. to 16°C (61°F), 75% RH, causing the product to dry out. It is dried to a weight loss of 20–40%, depending on the recipe; final moisture content is about 30–35%. This may take 3–7 weeks, depending mainly on diameter. A maturing period of six months or so may follow.
(5) Smoke may be applied before or during the drying stage.
(6) Mould growth may occur on the surface during the drying stage, accidentally or from cultures deliberately introduced. The mould is considered good for flavour and helpful to the drying process.
(7) The combination of lactic acid, low moisture content (= low water

activity) and the residual nitrite content gives long shelf-life ($\frac{1}{2}$–2 years or more) at ambient temperatures. The residual nitrite also forms and stabilises the intense red colour of nitroso-myoglobin.

The whole process requires careful control in order to manage exactly the hygienic conditions, temperature and humidity at every stage.

Corned beef

History

In the mid-19th century, the major industry of Argentina and Uruguay was the export of hides for leather making. The cattle carcasses were thrown away. In about 1847, Justus von Liebig invented a process for making meat extract by boiling lean meat with water, separating the solid residue and concentrating the aqueous portion by evaporation (50 parts meat give 1 part meat extract containing *c.* 18% moisture). This process was operated using the previously rejected meat; the solid residue after extraction was rejected. Shortly afterwards a process arose to salvage the boiled meat residue by mixing it with 'corns' or grains of saltpetre (sodium nitrate), canning it and giving a substantial heat process ($F_0 = 15$–20 or more). This was corned beef.

After the Second World War new processes arose to copy the earlier product without copying the whole process. Corned beef products are now made in Africa, Europe and elsewhere from unextracted meat, with more or less fat present, cured with nitrite or nitrite + nitrate, with variable proportions of added water, and canned. Salt and phosphates may be used in the curing process and the heat process may be mild, as appropriate to a luncheon meat.

Composition

South American corned beef, made by the original process as indicated above, has an apparent meat content of 120–125% by conventional analysis based on the nitrogen content. This is a result of the losses of water and soluble materials in the process of making meat extract. The fat content is fairly low (*c.* 15%). Other corned beefs may have much lower meat contents and include higher fat contents. From the nature of the processes used, the texture may be more like a luncheon meat than a traditional corned beef.

Technology

The factors governing yield and colour of the traditional product have not been studied outside the manufacturers and cannot be commented upon. It is evident that for both kinds of corned beef the conditions are quite different from those in most other meat products.

The factors involved in the manufacture of the modern products are closely similar to those governing a luncheon meat or a sausage made with beef. For the chopping technology see page 47).

10 Miscellaneous Meat Products

MINCED MEAT AND MEAT PREPARATIONS

For legal purposes in the EU the term 'Meat preparation' is applied to products consisting of uncooked meat with or without added foodstuffs, seasonings or additives but with no other treatment sufficient 'to modify the internal cellular structure of the meat and thus to cause the structure of the fresh meat to disappear'. Hamburgers with high meat content, British sausages and some of the products mentioned below fall into this category.

'Minced meat' refers to fresh meat which has been passed through a mincer; it does not include MRM.

'Lorne' (in Scotland) is beef sausage meat.

INJECTED MEATS

(a) Meat joints, poultry, poultry portions, etc. may be injected with solutions of salt, polyphosphate, flavouring agents, etc., with the following effects:

- The yield of uncooked product, i.e. the weight offered for sale, is increased.
- The yield on cooking is also increased, even if only water is injected (page 28); with salt and/or polyphosphate it is further increased (pages 29–30).
- Juiciness of the cooked product is therefore probably increased.

Note that under current regulations any additions of water above 5% require declaration on the label (see page 200).

(b) Poultry which is chilled in water or ice-water gains in weight. EU Regulations govern the hygiene of the process and limit the amount of water uptake for chickens, hens and cocks to 6%; lower figures apply for larger turkeys, etc.

(c) Meat or poultry may be injected with butter, oil, etc., to provide 'self-basting' during cooking and improvement in flavour, or with specially prepared broths, gravies, etc. These processes are generally protected by patents.

In all these cases note that the injection process can introduce contamination which may cause hazard or reduce shelf-life. Attention should be paid to:

- regular, frequent cleaning of equipment;
- temperature of injection solutions and equipment;
- possibilities of contamination of injection solutions, especially if solution is recycled or held for excessive periods. (Filtering, ultra-violet treatment, etc. may be useful.)

MEAT (INCLUDING POULTRY) WITH STUFFING

Stuffing is usually made from breadcrumbs, suet and suitable herbs; water is added immediately before use. For example:

Dry breadcrumbs	50
Shredded suet	45
Dried herbs	5
	100
plus added water, say	100–150

Stuffing may be introduced into the product, using, for example, a sausage filler.

Note the possibilities of introducing contamination via the stuffing, equipment, etc., especially if wet stuffing is delayed in process. The shelf-life of the product may be limited by the shelf-life of the wet stuffing.

CANNED MEATS

For canning processes see pages 112ff. For meat with gravy products of various kinds, there are no special problems. For luncheon meats, see page 165, and for corned beef, see page 168.

COATED AND BREADED PRODUCTS

This category may include:

- chicken portions, pork or lamb cutlets
- patties or 'fingers' made with comminuted meat
- rissoles, croquettes, etc.

Integrity of the product

It is usually necessary that the product being coated should remain fairly well intact by itself; the coating cannot be expected to hold a crumbly product together.

- Meat portions, cutlets, etc., present no problem.
- 'Fingers' and some other products may be prepared and sold in the frozen state to keep them intact.
- Rissoles, croquettes, etc. are usually made with potato, flour or other binders to ensure that the patties hold together before and after cooking.

Coating

Batter

This may be either:

- The main coating of the product, without any breadcrumb layer. Batter coats must have an appropriate viscosity to be applied in a layer of the required thickness, and are sometimes intended to 'puff' on cooking, giving a light, aerated layer. Basic composition will include flour and water, and baking powder. It may include milk, egg or other soluble protein to give 'body'. Or:
- An adhesive layer to hold a coating of breadcrumbs in place.

Breadcrumbs (rusk)

Proprietary brands are available. Texture and size distribution may be important. They may be coloured or flavoured.

Application

Combined battering and breading machines are usual. The product passes through on a conveyor; it is 'enrobed' in fluid batter, which falls in a thin film or curtain from a reservoir above the conveyor; it is then coated with dry breadcrumbs falling from a similar reservoir. Excess batter and crumbs are recirculated through suitable pumps. There may be devices to blow or shake excess loose material from the coated products, and to remove clumps from the recirculated crumbs.

Frying

The coating may be set in place by cooking at the time of manufacture, usually by frying. The conveyor from the battering/breading machine passes through a bath of hot edible oil.

Oil temperature is usually 180–190°C (355–375°F); appearance of smoke (200°C, 390°F or higher) indicates that the temperature is too high, causing deterioration of the oil.

The oil must be efficiently filtered to remove solid matter, which otherwise will hasten deterioration. Filters should be emptied and cleaned daily.

Under normal conditions of use, fresh oil is added regularly to replace oil absorbed by the product. This tends to maintain the average quality of the oil in the fryer. However, rancidity will increase with time and all the oil in the fryer must be replaced when necessary (e.g. if smoke is observed at lower than normal temperature or when the free fatty acid exceeds some specified limit, usually 2% as oleic acid). Intermittent use will shorten the average working life of the oil.

MEAT PIES

A variety of types of pie are made in the UK, as indicated in Table 10.1.

Table 10.1 Types of meat pie common in the UK

Type of pie	Filling	Pastry	Baking and consumption
Cold eating (e.g. 'pork pies')	Solid cf. sausages or luncheon meat + jelly filling	Short bottoms and tops, or boiled paste	Baked after making; eaten cold
Hot eating (e.g. meat and gravy pies)	Meat, etc., with gravy	Short bottoms; short or puff tops	Baked after making; eaten after reheating
Scotch pie	Meat, etc., with gravy	Boiled paste	
Frozen	Usually meat, etc., with gravy	Short bottoms; short or puff tops	Sold unbaked; baked before eating
Sausage rolls	Sausage meat or similar	Puff; sometimes short	Baked after making, eaten cold of after reheating, *or* unbaked, frozen, baked before eating
Vol-au-vents	Meat, etc., with sauce	Puff	Baked after making; eaten cold or after reheating
Pastries; Forfar bridies	Meat and vegetables with or without a thick sauce	Puff or formed around the filling	As sausage rolls

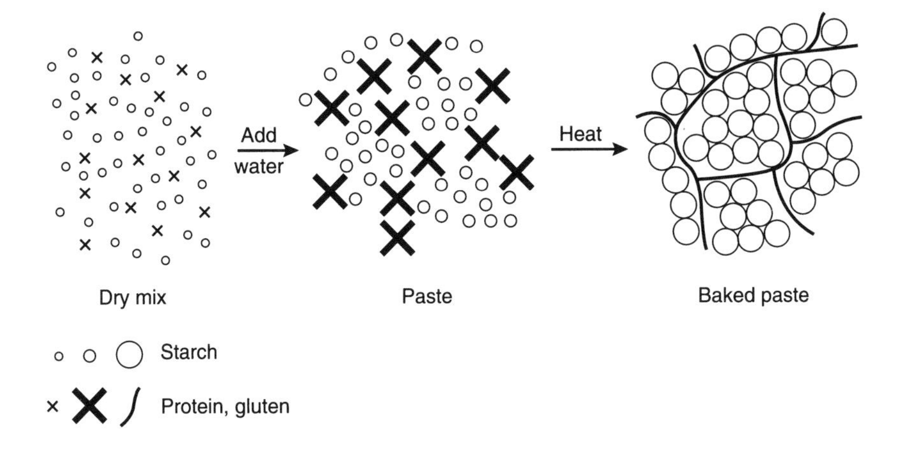

Fig. 10.1 Flour paste.

Pastry

Composition

Flour paste or **dough** comprises, essentially, flour + about 30% water. (+ some salt, for flavour only). The water is taken up by:

- the protein (gluten) to form a tough, sticky matrix
- the starch granules, to some extent.

On heating, the gluten structure loses water, coagulates and becomes hard and rigid; the starch granules absorb some of this water, gelatinise and remain soft, embedded in the gluten. (See Fig. 10.1.)

Fat breaks up the gluten structure.

Short pastry

Short pastry is made by mixing in the fat finely, giving a 'short' paste when moist, and crumbly pastry when dry (see Fig. 10.2).

Puff pastry

Puff pastry is made by laminating fat with a short paste or a paste with very low fat content; this is done by folding and rolling out several times (see Fig. 10.3). The laminating fat may be added at the beginning of the rolling stage or may be rough-mixed with the pastry beforehand (Scotch method). The

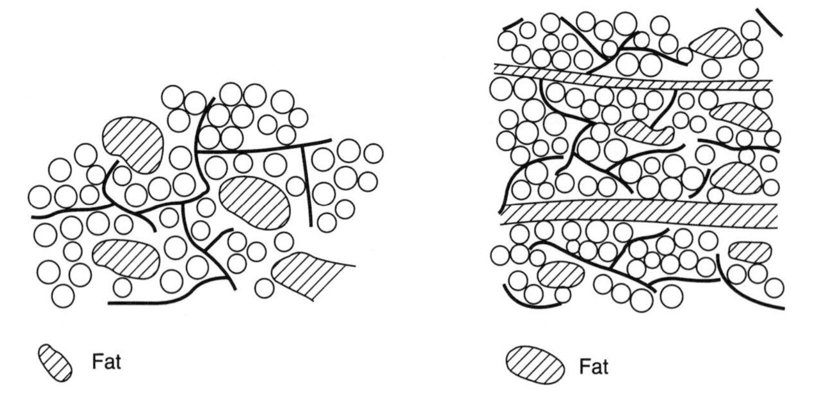

Fig. 10.2 Short pastry.

Fig. 10.3 Puff pastry.

fat (lard or bakery margarine) must be reasonably plastic: not too hard or it will not form layers easily; not too soft or the layers will break.

On baking, steam accumulates at the sheets of fat and blows the pastry apart in layers. The more layers and the thinner the paste between, the greater the 'puff'.

Table 10.2 Relevant temperatures in pie cooking

Temperature °C	°F	Starch	Gluten
5–10	40–50		Rapid hydration to stringy pastes
52	125	Swelling begins	
62	143		Coagulation begins; some water loss
70	158	Bursting of grains; gelatinisation begins	
74	165		Coagulation irreversible; large water loss
93	200	Gelatinisation complete	
Baked at: 150	300		Bakes to hard compact mass
Baked at: 230	450		Rapid evolution of water vapour gives crisp friable mass

Typical pastry recipes

	Short	*Puff*
Flour	70	70
Fat	30	35–50 (pastes 25–35, laminate 10–20)
Salt	$1\frac{1}{2}$	$1\frac{1}{2}$
Water	22	22–30

For general information on pastry making, consult a textbook of baking technology, e.g. Street (1991).

Hot pastes are made with hot or boiling water, with fat as for short pastry. This gelatinises some of the starch at the mixing stage, making the paste stiffer. This is traditionally used for large, hand-moulded pies.

Relaxation (resting)

Gluten pastes are plastic and easily stretched, shaped, moulded, etc., but immediately after stretching or 'working' in any way they are also very elastic; thus:

- If released (e.g. from stretching), they tend to spring back, distort, and resist further changes of shape.
- If baked in this condition, they shrink and become tough.

On resting, the elasticity disappears; the paste becomes plastic again and can be moulded, etc., as before. If baked in this condition, the paste does not shrink or become tough. Therefore, to ensure absence of distortion in shaped or cut goods, and absence of toughness on baking, pastry should be rested after it has been worked: e.g. $\frac{1}{2}$–1 h after rolling out; $\frac{1}{2}$ h between forming pies and baking.

Use of a relaxing agent (cysteine, sodium bisulphite, various soya preparations) may reduce the need for resting but may also give a softer pastry.

Re-use of scrap paste

Large amounts of scrap are produced in all cutting and forming operations, e.g. cutting circles from a rectangular sheet gives about 25% waste. The scrap is normally incorporated into fresh paste. However, note the following points:

- The scrap should be allowed to rest, or used with ascorbic acid in a high-speed mixer.
- It should be used for short paste, not puff.
- Scrap is less fresh than the new paste and has higher microbial counts.

Especially in warm weather, continued incorporation of scrap will lead to souring of the whole. Check scrap quality (e.g. pH value should not be below 6.4) and when too much deteriorated reject it (e.g. bake, grind and use as a cereal filler). This is usually necessary once or twice a week. Do not hold over a weekend.

Pie making

A typical line system comprises:

- a device for cutting lids from sheet pastry
- moulds and dies for forming ('blocking') bottoms, from balls of dough supplied by a dough divider, or from sheet pastry
- filler, in two stages if necessary, e.g. for meat pieces and sauce
- a device for applying and sealing lids to the bottoms
- glazing, if required.

Most standard pie machines combine several of these stages.

Pie fillings

Solid fillings for cold-eating pies and sausage rolls

The technology of manufacture is as for sausage meat (page 47) or luncheon meat (page 165). Sodium nitrite (e.g. 100 ppm) is included in some recipes to give a pink colour in the cooked pie. **Jelly** (if needed) requires, for example, 6% gelatin (lime or acid processed) or 4% gelatin (lime processed only) + 2% agar. The following procedure should be followed:

- Keep solution at 80–90°C (176–190°F) before injection; lower temperatures give risk of microbial growth; boiling the solution causes reduction of gel strength.
- Inject hot solution into the pies just after baking, when the pastry has cooled. Allow to cool further until the jelly sets, before handling; otherwise there is risk of absorption of jelly by the pastry, therefore sogginess.
- Use of agar raises the setting temperature and allows pies to be handled sooner.
- Ideally, pie centre temperature should be 24°C (74°F) during jellying, and cooling of the pie should be completed immediately afterwards.

Meat-with-gravy fillings for hot-eating pies, pasties, etc.

Diced or pre-ground meats are cooked together, usually in open-topped cooking pans, with the bulk of the water and seasoning. Vegetables, and less tough meats, e.g. kidneys, are then added. After further cooking, starch additives are used to effect thickening.

Adequate cooking of onions and flour is essential in order to avoid rapid spoilage of the cooled pie filling, particularly if this has to be stored before use.

Handle and fill at temperatures not less than 72°C (160°F). Keep agitated to avoid separation of the components.

Baking

Requirements (see Fig. 10.4)

To bake the pastry shell the following approximate oven temperatures are suggested:

- For short pastry start at 148°C (300°F); rise to 177°C (350°F).
- For puff pastry start at 232°C (450°F); rise to 246°C (475°F).

See page 176 for changes taking place.

To heat the filling, to cook and sterilise it a centre temperature of 85°C (180°F) is adequate. Times will depend on pie size.

Note that pies that have been frozen or stored under chill conditions require longer baking times unless they are allowed to warm up first.

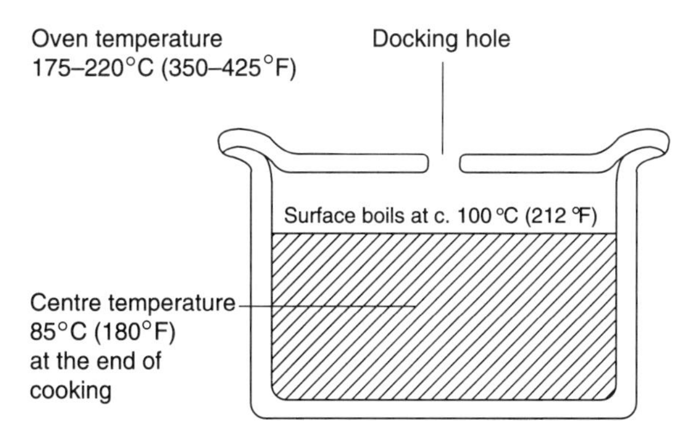

Fig. 10.4 Baked pie temperature.

Boiling out

If the pie filling at the surface boils during baking and forms a stable foam, bubbles and liquid may rise through the docking hole. If the docking hole is blocked, the lid may be lifted off when the contents boil.

Possible causes and preventive measures are:

- Headspace too small: do not overfill.
- Filling too thick:
  - reduce viscosity, etc., of filling to reduce tendency to foam
  - change starches, flours or other thickeners in gravy
  - reduce or greatly increase pre-cooking of gravy to render it less thick.
- Fat loss may sometimes be associated with increased tendency to foam: see Chapter 2 for general methods of reducing fat loss.
- Too fast heating at surface of filling: reduce oven temperature.

Cooling

Rapid cooling is essential for good keeping quality. Vacuum cooling is a very effective process but involves high capital cost. Cooling is usually achieved with circulating air. Problems encountered are as follows.

Hygiene

The air in a pie department is likely to be contaminated, in particular with mould derived from the flour. It is best to provide a separate cooling area, if possible, to minimise contamination from this source. Ideally, use filtered air.

Condensation

This readily leads to sogginess and to mould growth and spoilage. See page 182 for detailed descriptions. Most likely causes are:

- insufficient cooling before wrapping – leads to 'fog' on the inside of wrapping film
- insufficient cooling and/or temperature fluctuations after production is complete – leads to damp undersurfaces of lids, causing mould growth inside the pie
- mould growth on vulnerable parts of the factory structure – provides a source of contamination of the product.

Staling

In baked pie pastry the crisp outer layer becomes softer by transfer of moisture from the inner layers. This staling is delayed by low storage temperatures because the moisture transfer is slower.

However, a second form of staling is maximal at temperatures around

0°C (32°F). This is due to crystallisation of the gelatinised starch, causing a hard dry texture as in stale bread.

Freezing

Unbaked pies

No special problems arise in this case.

Baked pies

Freeze and thaw rapidly. No staling occurs when the moisture in the pastry is frozen, but any condensation of moisture from the air (e.g. during thawing) may cause softening.

Thawing

It is difficult to ensure minimum time at +1 to +7°C (32.5° to 45°F) during thawing. Controlled thawing in warmed air may be helpful, or re-heating where this is appropriate. For these reasons, baked pies, etc., are not commonly frozen.

Storage and transport

Unbaked, unfrozen pies

No special problems arise. Temperatures should be as low as practicable. (Glazed pies may become crazed if the temperature is too low.)

Unbaked, frozen pies

No special problems arise. Normal frozen handling applies.

Baked pies

Temperature has to be a compromise. For the filling it should be as low as practicable; for the pastry, it should not be below 7°C (45°F) (see above). A temperature of 7°–10°C (45–50°F) is considered a suitable compromise.

Mould growth

Conditions for the growth of mould inside pies are shown in Fig. 10.5.

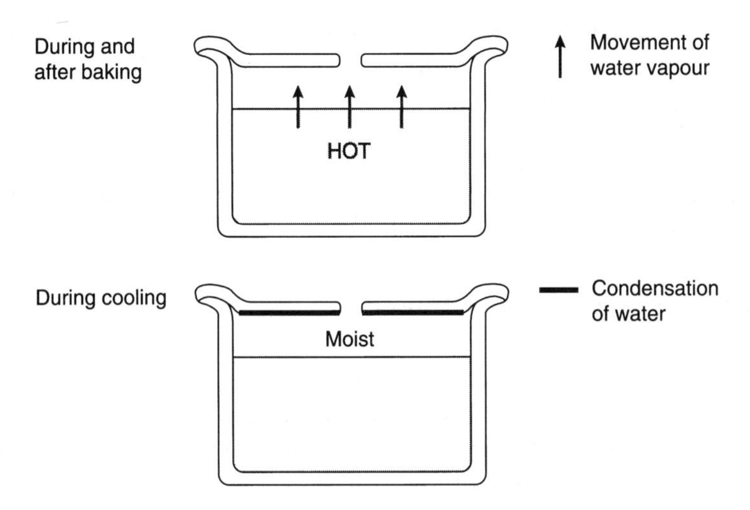

Fig. 10.5 Conditions for mould growth inside a pie.

If cooling is incomplete,

- more condensation will take place
- mould will grow where the pastry has been moistened by condensation.

Mould growth in pies is a common cause of rejection; it should always be taken as an indication of mishandling by improper temperature control.

PRODUCTS MADE FROM TRIMMINGS, OFFAL, ETC.

General

Various products are made from traditional recipes using edible offals and meat trimmings. Some of them are highly regarded specialities, though usually in quite localised markets. In many places, however, consumer acceptability of offal products is low and there may be little scope for products of this kind.

From the point of view of the manufacturer (whether large- or small-scale) it is usually very desirable to have at least one product which can be formulated to include various materials unsuitable for other processes but otherwise edible and wholesome, such as meat trimmings, rind and gristle, damaged or misshapen products, excess pastry from pie manufacture, etc. The recipe need not consist entirely of such materials; ideally it should be possible to include them in variable amounts, within fairly wide limits, without significantly affecting the properties and quality of the final product.

The recipes quoted in this section are intended only to give an indication of a satisfactory, simple way of making each kind of product. Wide variations are possible and will be encountered in practice.

Note that there are some legal restrictions upon the offals which may be used, especially following the BSE problem. In the UK at the time of writing, certain offals of cattle and sheep are effectively excluded from use in meat products for human consumption.

Black pudding (blood sausage)

A typical recipe is given in Table 10.3. The product uses slaughterhouse blood, usually in places where slaughterhouse and processing plant are close together (stored or transported blood is not usually used) and where there is a traditional market. In the UK this is mainly in the Midlands and the North.

The product is firmly held together by the heat-set blood. Increase in the amount of cereal filler, over the quantity given above, is possible. In the UK the product was once commonly coloured with black PN or Brilliant Black (E151) but these are not now permitted in the EU.

Table 10.3 Recipe for black pudding

	%
Blood	48
Diced fat (beef suet or pork flare fat)	24
Cooked barley, flour or oatmeal	18[1,2]
Salt	5
Onion	3
Spices	2
	100

Notes

1. Approximately dry cereal 5, water 13.
2. Some scrap pie material, etc., may be included.

Process:
Mix the ingredients, fill into wide casings; cook in a hot water bath at 82°C (180°F) for 40 min.

Problems

Handling of raw blood

Anticoagulant may be used; if not, clotting will occur, giving material which can be inconvenient to handle at the mixing stage.

Undercooking

This is an indication of poor factory control, and may lead to rapid souring.

Mould growth

This is possibly due to defective raw materials, but is more likely to be due to condensation effects after cooking (pages 80, 182).

Colour

The colour should be very dark brown or black, with white pieces of fat.

- Grey-green – excessive air in the product.
- Red – possibly undercooking, *but* the centre of the product may be red even after normal cooking. This is probably a version of the red colour found at the centre of some cooked uncured meats, due to a reduced form of the cooked pigment (see page 69). In some cases it appears to be associated with the use of anticoagulant. In the UK the colour Black PN is used to overcome this.

Fat loss ('fat capping') on cooking

Fat loss is probably the result of:

- poor dicing and excessive damage of the fat
- incomplete mixing; individual fat dice should be fully enclosed in the blood mixture.

Products made mainly with offal

Haggis

A typical recipe is given in Table 10.4. The mixture is filled into a sheep's stomach (traditional) or other casing and cooked in boiling water, e.g. 50 min for 1 lb size. There is no legal requirement to declare a meat content (UK).

Problems

These include souring and mould growth (see under black pudding, above).

Table 10.4 Recipe for haggis

Pluck * and/or other offals	53
Suet, fat trimmings, etc.	17
Oatmeal	6
Water or stock	20
Salt and pepper	2
Onion	2

* Pluck = trachea + lungs + heart + spleen (see page 22).

Andouillette

This is the traditional French method of using pigs' stomachs and intestines. Wide variations in recipe, cooking and presentation can be found, but essentially a mixture of stomachs and intestines (33–45% stomachs, according to availability), with or without pre-cooking, is minced or chopped, then cooked in water until reasonably tender (typically for 4 hours at 90°C, 194°F).

For a restaurant dish, the offals may be cooked and then served in a liquid containing salt, spices and possibly thickeners.

The manufactured product is usually filled into sausage casings, with added nitrited salt, before cooking in the casings.

Products made with trimmings and some offal

Brawn, Bath chaps

These products are made by boiling in water:

- bones
- rind, gristle, etc.
- head meat, ears, feet, etc.
- other trimmings high in connective tissue, low in fat. Bath chaps are made mainly from boned-out pig cheeks.

Enough water should be used to give a stock with high gelatin content; e.g. use two to three times as much water as solid matter and boil down to about half the original volume. Vinegar may be added to improve hydrolysis. Gelatin may be added (e.g. 1% in final product) to improve jelly strength. See page 81 for problems associated with using hot gelatin solutions.

The products may also contain cooked meat, rusk, etc., *or* may be clarified with egg white or blood plasma to give a clear jelly or 'aspic' surrounding pieces of cooked meat, etc. Such products are usually cured by addition of 150–200 ppm sodium nitrite to the cooked meat.

The UK Regulations do not require a meat content declaration.

Table 10.5 Typical recipes for haslet, saveloys and faggots

	Haslet %	Saveloy, polony %	Faggot, savoury duck %
Meat trimmings, MRM, Offal[2]	75[1]	70	70
Rusk	9	7	18[3] (rusk and pie material)
Pie material	–	–	
Water	12.5	18	28
Salt	2.5	4.5	2
Herbs, spices	1.5	0.5[4]	2[5]
Process	Coarse chop. Form into cubes, etc. Cover with caul fat; roast, e.g. 150°C, 300°F	Coarse chop. Fill into sausage casing. Smoke if required. Cook in moist air or hot water, e.g. 15 min at 80°C, 175°F	Chop, form into balls. Either: (a) cover with caul fat, roast, e.g. at 150°C, 300°F, or (b) pack in trays, etc. with gravy, cook in oven. In either case, cook to centre temperature 70–75°C, 160–168°F or higher

Notes
1. At least 35–50% lean meat.
2. Pork 'white' offals; may or may not be included in the recipe.
3. Variable proportions of rusk and scrap pie material.
4. Includes smoke flavour if the product is not to be smoked.
5. Normally onion.

Haslet

Haslet is a speciality of the English Midlands. The proportion of meat is fairly high and the recipe is close to a sausage formulation; offals may or may not be included. It is commonly sold as a cooked product with short shelf life, in cubes or similar pieces, unpackaged.

Saveloy or polony

This has a lower meat content than haslet and may have a higher offal content. It is filled into casings and presented in the form of sausage.

Faggot or savoury duck

This is of similar composition to saveloy, and is usually presented in balls of 20–30 g, either roasted or cooked in gravy.

Typical recipes are given in Table 10.5. In all cases the list of ingredients and the true meat content (which may include MRM but not offal) need to be declared on labels.

The integrity of the manufactured products depends partly on binding of the lean meat, usually to a greater extent on the plasticity of the cooked cereal portion.

Problems which may be encountered, and suggested solutions, include:

- poor binding: ensure cooking temperature is high enough;
- fat cook-out: increase lean meat content; use a heat-setting starch;
- souring: check the microbial quality of ingredients; check cooking and cooling temperatures
- mould: check for condensation, etc. (see pages 80, 182).

Other products

Scotch egg

This comprises a boiled egg in a sausage meat or similar casing, with batter and breadcrumb coating. The approximate proportions are:

boiled egg	28
casing	58
coating	14

The casing is usually made by simple sausage technology (page 47). For the breadcrumb coating, see page 173.

Problems

Blackening of the egg (see Fig. 10.6). Hydrogen sulphide, H_2S, is released from the egg white protein when it is cooked. There is a high concentration of iron in egg yolk. A black layer of iron sulphide, FeS, is formed at the boundary. This is harmless but unsightly.

Treatments include:

- Rapid and thorough cooling after boiling (use cold water or ice-water), which sets the egg white and reduces the movement of H_2S.
- Dipping the boiled and shelled eggs in hydrogen peroxide overnight. This oxidises the FeS to $FeSO_4$ (iron sulphate), which is almost colourless. Residual peroxide is not usually to be found in the end product.
- Dipping in citric acid or sodium citrate. This complexes the iron and prevents reaction. Note that during prolonged use the composition of the dip must be maintained.

Application of sausage meat casing. This is often done by hand. A Rheon machine is also suitable

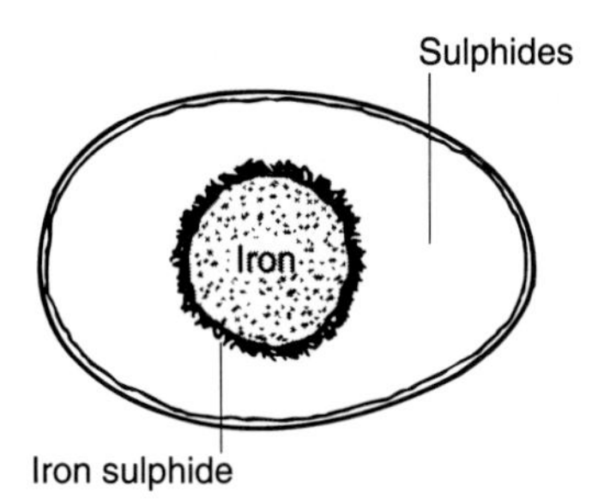

Fig. 10.6 Blackening of boiled egg.

Rissoles, croquettes, etc.

These are made with cooked meat (e.g. 10–20%), usually minced, with rusk, starch, potato, etc. They are moulded into shape and usually coated with breadcrumbs (page 173). Product integrity depends on the binding of the cooked starches.

If meat is mentioned in the product name, meat content will need to be declared. (Note that the declaration refers to the proportion of uncooked meat used in the production of 100 parts of final product. If this is not known, assume 100 parts cooked meat comes from 125 parts raw meat.)

White pudding

This is a local speciality in Ireland, Scotland and some parts of England. It consists of water-soaked oatmeal and beef suet, filled into casings and cooked in boiling water. The product is held together by the oatmeal.

A typical Scottish recipe is as follows:

Soaked oatmeal	48
Beef suet	44
Leeks	6
Salt	2

Sandwiches, ready meals, etc.

There is a wide range of short-shelf-life food products in which meat or a meat product is one of the ingredients but not necessarily the most important. The other ingredients present – e.g. the bread and the butter in a sandwich, vegetables and gravy in a ready meal – will have different properties, different preparation processes, different shelf life, and different storage requirements from one another and from those of the meat. The formulation of such complex products, and the specification and control of their packaging and the conditions of storage, transport and display, require considerable skill in a wide area of food technology beyond that of the basic meat technology outlined in this handbook. Meat technologists considering entry into this area are strongly advised to equip themselves first with the necessary understanding or else to find and work closely with colleagues already equipped.

11 Controls – Manufacturing, Commercial and Legal

MANUFACTURING CONTROLS

It should go without saying that every manufacturing operation must be properly controlled, in order to make products:

- of the required composition, and
- of consistent quality.

'Quality' should be taken to mean:

- in conformance with specification
- within agreed tolerances of variability within and between batches.

Specifications should be agreed in advance:

- by all relevant departments within the company and, as appropriate
- by the intended customer(s) (who may sometimes indeed lay down their own specifications).

Good manufacturing practice (GMP)

To achieve all the above it is essential to follow good manufacturing practice, which requires:

> '... that every aspect of the manufacture is fully specified in advance and that all the resources and facilities specified – namely
>
> - measures and precautions at critical control points based on hazard analysis;
> - adequate premises and space;
> - correct and adequately maintained equipment;
> - appropriately trained people;
> - correct raw materials and packaging materials;
> - appropriate storage and transport facilities;
> - written operational procedures and cleaning schedules;
> - appropriate management and supervision; and
> - adequate technical, administrative and maintenance services

are in fact provided, in the right quantities, at the right times and places, and are utilised as intended.'

Quoted with permission from IFST *Guide to Good Manufacturing Practice*, obtainable from IFST, 5 Cambridge Court, 210 Shepherd's Bush Road, London W6 7NJ.

The IFST Guide goes on to elaborate the steps which need to be taken by management, especially including higher management, to ensure that all these requirements are met.

At least the following need to be in place and routinely followed:

- written recipes and processing instructions for each product, immediately available at the points of use;
- methods and instructions for cleaning and sanitising of equipment and premises, similarly available at the relevant points of use;
- full (but simple) records of all operations carried out;
- a system of **traceability**, by which the date of acquisition, the source and so far as possible the full previous history of every meat carcass, batch of manufacturing meat or other raw material are clearly established and recorded (in this connection note the requirements of the Beef Assurance Schemes in the UK and Ireland);
- a **quality manual** in which are clearly set out:
  - the precise procedures and the personnel responsible for each operation from acquisition of raw materials to disposal of the products and of all wastes
  - procedures for the identification and correction of mistakes and the disposal of defective materials (this 'quality control' function may be exercised by a quality control department largely independent of manufacturing or – better – with controls applied directly by the manufacturing staff, the test procedures being supplied and the results monitored by a quality assurance function)
  - a system of regular internal audits to check that all procedures are properly followed and to ensure corrective action where they are not.

If all these things are properly done, there should be little need for special efforts or precautions to meet the demands of particular customers or official inspectors, beyond co-operating in any meetings, inspections or independent audits which they intend to make, and of course dealing with the associated paperwork.

Under present circumstances in the UK the provision and maintenance of all the necessary facilities and the proper application of appropriate procedures, together with documentation to confirm that all relevant points are covered, will be critically important if any product should be found or alleged to be contaminated or spoiled or a source of food poisoning, causing harm and leading to a threat of legal prosecution. The manufacturer will not be prosecuted, or can defend himself is he is prosecuted, if he can show

- that the fault was not caused by him, or
- that he had foreseen the possibility of the occurrence of such a fault **and had taken reasonable care to prevent it** – as evidenced by the written procedures and the quality manual,
- and that he had **shown due diligence** in applying the procedures – as evidenced from the written and audited records.

Codes of Practice

In arranging and doing the things required by good manufacturing practice, there are various relevant Codes of Practice which should be followed.

A Code of Practice is a document agreed between manufacturers and the legal authorities and sets out 'the recognised procedures, practices and standards for a food operator ... in the general interests of efficiency, safety, good management and the production of high quality food products'.

We can consider them to be of three kinds, as follows.

Meat industry Codes of Practice

These set out methods of manufacture and standards for finished meat products, agreed by industry representatives as reasonable, attainable in practice and commercially viable. 'Industry representatives' include, most particularly, the British Meat Manufacturers' Association (BMMA) in the UK and the Liaison Centre for the Meat Processing Industry (CLITRAVI) in the EU. In the UK other bodies such as the Meat and Livestock Commission (MLC) or the Local Authorities Committee on Trading Standards (LACOTS) are involved in some cases. Though these Codes are not legally enforceable, the enforcement authorities will always regard any deviation from their provisions at least with suspicion, as likely indications of malpractice.

Food industry Codes of Practice

These are drawn up by other bodies than the meat industry, for instance the Food and Drink Federation, and other national and international bodies such as the Ministry of Agriculture Fisheries and Food (MAFF) and the World Health Organisation (WHO). They tend to be wide in scope.

Statutory Codes of Practice

These are published by the UK Government under the Food Safety Act 1990 (see later). They give guidance to the relevant authorities on all aspects of enforcement of the Act. Meat products manufacturers will be wise to make themselves aware in advance of those features that will be applied whenever their own operations are to be inspected or challenged in any way.

A full list of Codes relevant to products in this handbook is given in the IFST publication *Listing of Codes of Practice Applicable to Foods* (1993): see section B5, Meat products. Codes, Standards and Guidelines produced by the BMMA are available from the BMMA, 11/12 Buckingham Gate, London SW1E 6LB.

COMMERCIAL CONTROLS

Cost control and recipe control

It is obviously essential that in any manufacturing operation the costs of materials, labour, services and capital must be held within profitable bounds. This subject, however, lies mostly outside the scope of this handbook.

Note, however, the contribution which computerised least-cost formulation programs can make to minimising ingredient costs. In these programs, conventional values for the fat and water contents, 'bind' values and other properties of a range of meat cuts (as in Table 8.2) are set against their current prices, to compute the most economical recipe for a given product and indicate the quantities of each material which should be purchased for, say, a week's production.

Controls exercised by retail customers

A manufacturing business which supplies meat products to a supermarket or other retailer customer may expect to receive direct attention from that customer who will wish to be assured that

- product made for them is in conformance with the specifications agreed at the outset, whether dictated by the customer or proposed by the supplier;
- 'due diligence' is shown in the observance of all the other requirements of GMP.

LEGAL CONTROLS

Finally, it is obvious that every meat product should be manufactured at all times in compliance with the law. This should present little problem if the product is made with careful attention to the technology as set out in the preceding pages and in all respects according to good manufacturing practice.

Of course, as the BSE crisis clearly demonstrated, new and unexpected requirements that are not always technologically justified may be imposed at almost any time in response to public or political pressure; no account can be taken here of any future developments of such kinds.

A brief summary of the main legal provisions in operation at present in the EU and the UK follows.

General food law

In the European Union

The general food law of all member states is governed by a number of directives and regulations. Directives are laws which lay down general standards and 'direct' each country of the EU to frame or amend its internal legislation to conform with those standards. Regulations are intended to be applied directly by all the countries.

Directives concerned with foods in general include:

- Council Directive 93/43/EEC, on the hygiene of foodstuffs
- Council Directives 79/112/EEC and 89/398/EEC, on the labelling, presentation and advertising of foodstuffs

Directives and regulations directly concerned with meat products include:

- Council Directive 92/5/EEC, on public health problems of meat products
- Council Directive 94/65/EC, on public health aspects of minced meat and meat preparations
- EC Regulation 1538/91 (as amended), on marketing standards (including water content) of poultry meat

In the UK

Food legislation in the UK was almost all already in place and operational before the EU embarked on the long process of harmonisation of the laws of the individual member states. Now the harmonisation process is virtually complete and the UK law has been amended wherever necessary to comply with the EU standards.

The principal legislation covering all food manufacturing operations is contained in:

The Food Safety Act 1990
The Consumer Protection Act 1987
The Weights and Measures Act 1985

and in a number of subsidiary regulations:

The Materials and Articles in Contact with Food Regulations 1987, as amended
The Food Safety (General Hygiene) Regulations 1995, as amended
The Miscellaneous Food Additives Regulations 1995, as amended
The Colours in Food Regulations 1995
The Temperature Control Regulations 1995
The Food Labelling Regulations 1996, as amended
The Contaminants in Food Regulations 1997, as amended
The Plastic Materials and Articles in Contact with Food Regulations 1998

Regulations particular to meat and meat products include:

The Meat Products and Spreadable Fish Products Regulations 1984, as amended
EC Regulation 1538/91 (relating to water content of poultry)
The Meat Products (Hygiene) Regulations 1994, as amended
The Minced Meat and Meat Preparations (Hygiene) Regulations 1995
The Fresh Meat (Beef Controls) (No. 2) Regulations (1996)*
The Animals and Animal Products (Examination for Residues and Maximum Residue Limits) Regulations 1997
The Specified Risk Material Regulations 1997, as amended*
The Beef Labelling (Enforcement) Regulations 1998

'QUID' labelling – under which the percentage declaration of the quantity of an ingredient is required if the ingredient appears in the name of the food, is emphasised in words, pictures or graphics, or is essential to characterise the food – comes into force in the UK early in 2000, under the 1998 Amendment to the Food Labelling Regulations. ('QUID' refers to the EC Quantity Indication Directive.)

Under the Food Safety Act 1990 there are also statutory Codes of Practice. These are mentioned on page 194.

* Sets out restrictions due to BSE.

ANALYTICAL CONSIDERATIONS

The relative amounts of the major components of meat are shown in the diagram below.

'Meat'	consists of:	with chemical composition:
Lean meat =	Muscle 75–97% (including CT *c.* 9%) + Other CT 0–15% + Fat 3–10%	Protein 23% Water 77% Fat 3–10%
Fatty tissue or 'fat' =	Fat (lipid) *c.* 90% + CT *c.* 10%	Fat 90% Protein 2.2% Water 7.8%

CT = connective tissue.

The water/protein ratio in muscle (77/23 = 3.35) is extremely constant. In lean meat or in whole meat including fat, the figure is virtually the same, being little affected by the proportion of connective tissue or the proportion of fat.

For many purposes it is convenient to take the water content of lean meat as 75%, which makes some allowance for the fat present.

For whole meat, where no water or protein has been added or removed, the fat content can be calculated from either the water content or the protein content:

$$\text{Fat} = 100 - (\text{water} \times 4/3)$$

or $$\text{Fat} = 100 - (\text{protein} \times 100/23).$$

Meat products and other complex mixtures are normally analysed by way of the nitrogen content. There are two different approaches to this. In the UK, Ireland and some British Commonwealth countries the nitrogen content is used to calculate the 'lean meat content' of the product; in the other EU countries, the USA and elsewhere, the nitrogen/water ratio of the meat product or another closely related factor is compared with an agreed factor for genuine meat.

Estimation of meat content

The formula for calculation of the 'apparent lean meat content' is:

$$\text{Lean meat content (\%)} = \frac{\text{Nitrogen content (\%)}}{\text{Nitrogen factor}}$$

using the appropriate Nitrogen factor selected from Table 11.1.

Table 11.1 Nitrogen factors for meat content analysis

Pork	3.45	Liver, ox	3.45
Beef	3.55	Liver, pig	3.65
Veal	3.35	Liver, unknown origin	3.55
Chicken, breast	3.9	Kidney	2.7
Chicken, dark meat	3.6	Tongue	3.0
Chicken, whole carcass	3.7	Blood	3.2
Turkey, breast	3.9		
Turkey, dark meat	3.5		
Turkey, whole carcass	3.65		

The pork and beef factors above are average values (per cent nitrogen on the fat-free basis) for all cuts of meat from the animal in question. They may be incorrect for particular cuts whose composition differs markedly from the average, as is the case with many of the cuts used for manufacturing.

For pork, the following factors are recommended for use when the individual cut is known:

	Lean and subcutaneous fat	*Lean, rind and subcutaneous fat*
Collar	3.35	3.50
Hand	3.35	3.60
Middle cuts	3.50	3.75
Rib belly	3.45	3.70
Rump belly	3.45	3.70
Rib loin	3.60	3.80
Rump loin	3.60	3.80
Leg	3.45	3.60
Whole carcass	3.45	3.60

It is essential to make allowance for any nitrogen from non-meat sources present in the meat product.

- The Stubbs and More calculation makes allowance for the nitrogen in the bread or rusk content of British fresh sausages by estimating the starch content, assuming that the starch is associated with 2% nitrogen and deducting that amount from the total nitrogen.
- If milk or whey proteins are present, they may be detected and corrected for via an analysis of the lactose content.
- Soya flour may be detected and sometimes estimated microscopically; soya isolate may be estimated serologically (in unheated products only) or by DNA methods; the appropriate adjustment must then be made to the total nitrogen content.

The apparent total meat content is then given by adding the fat content to the apparent lean meat content.

Connective tissue

It is normally assumed that 10% of lean meat consists of connective tissue. Where pork meat is used in a product, then rind from the pork may be present in addition. Since the proportion of rind on the pig carcass is taken as 10% an additional connective tissue content up to10% of the total pork meat (lean plus fat) in a product is allowable. In an analysis of a meat product it may be necessary to check whether the connective tissue content is within these limits.

Connective tissue content (approximately equal to collagen content) can be estimated from the hydroxyproline content. Conventional factors are:

Wet connective tissue = hydroxyproline % × 37
Dry connective tissue = hydroxyproline % × 8

Where the connective tissue content is high, its effect may be allowed for by calculating a muscle protein content separate from the connective tissue. Use the following:

Collagen	= N% × 5.55
Dry connective tissue	= HyP% × 8
Nitrogen due to connective tissue	= HyP% × 8/5.55
	= 1.42 × HyP%
Nitrogen due to muscle protein	= N% − 1.42 × HyP%
Therefore muscle protein	= 6.25 (N − 1.42 × HyP%)

where: N = total nitrogen %; HyP = hydroxyproline %.

This corresponds with the BEFFE factor used in Germany. (BEFFE = Bindesgewebeeiweiss-frei-Fleischeiweiss = connective-tissue-protein-free meat-protein.)

Control via water content

The water/protein ratio may be used as an index of added water in meat or meat products. If other proteins are present, they must be allowed for as indicated above.

A closely related factor is the **Feder number**:

$$\text{Feder No.} = \frac{\text{Water\%}}{\text{Organic non-fat\%}}$$

where

$$\text{Organic non-fat\%} = 100 - \text{Fat\%} + \text{Ash\%} + \text{Water\%}$$

For control purposes in France and many other countries a maximum ratio of 4:1 is taken as the practical limit. In the USA the factor is 4.0 (Those limits are actually quite generous: a ratio of exactly 4.1 could be given by a mixture of 25 parts of water with 100 parts lean meat.)

In the UK the **added water** content of a meat product is calculated as

100% – (meat content% + other added solids%)

Regulations in the UK require that for meat products with high contents of water or curing brine the additional water over certain conventional values must be declared. For full details of the requirements, the calculations and the form of declarations required, see the Meat Products and Spreadable Fish Products Regulations 1984.

Identification of meat species

There is considerable interest from time to time in the detection of other meat species if these are fraudulently mixed with manufacturing meat, and of non-meat proteins incorporated into meat products.

Methods used include the following:

- Species-specific immunological tests. These have also been used for the detection of soya and other foreign proteins in meat products. Earlier versions of the tests did not work well on cooked or heat-treated material but this difficulty has been overcome in newer versions.
- DNA probes are becoming more widely available. They are completely species-specific and are not affected by heat treatment.

References and Bibiography

Advisory Committee on the Microbiological Safety of Food (1992) Report on Vacuum Packaging and Associated Processes. HMSO, London.

BMMA (1996) Standard for the Production of Bacon and Bacon Joints. British Meat Manufacturers' Association, London.

CLITRAVI (1997) European Standard for Mechanically Separated Meat, MSM/96/4. Centre de Liaison des Industries Transformatices de Viande de l'UE, Brussels.

Evans, G.G. & Ranken, M.D. (1975) Fat cooking losses from non-emulsified meat products. *J. Fd Technol.* **10**, 63–71.

Evans, G.G., Ranken, M.D. & Wood, J.M. (1975) Curing of Meat. Brit. Pat. No 1375700.

Food Micromodel (a computer program showing growth curves of pathogenic and food spoilage micro-organisms under various conditions). Available from Leatherhead Food Research Association, Leatherhead.

Gerrard, F. (1977) *Sausage and Small Goods Production*, 6th edn. Northwood Publications, London.

Harrigan, W.F. & Park, R.W.A. (1991) *Making Safe Food: A Management Guide for Microbiological Quality*. Academic Press, London.

Hersom, A.C. & Hulland, E.D. (1980) *Canned Foods*, 7th edn. Churchill Livingstone, Edinburgh.

Hoogenkamp, H. (1979) *Practical Applications of Milk Proteins in Meat Products.* DMV, Veghel.

IFST (1990) *Guidelines for the Handling of Chilled Foods*, 2nd edn. Institute of Food Science and Technology (UK), London.

IFST (1998) *Good Manufacturing Practice: A Guide to its Responsible Management*, 4th edn. Institute of Food Science and Technology (UK), London.

Ingram, M. & Kitchell, A.G. (1967) Salt as a preservative. *J. Fd Technol.* **2**, 1.

Larousse, J. & Brown, B.E. (1997) *Food Canning Technology*. Wiley–VCH, New York.

Lawrie, R.A. (1998) *Lawrie's Meat Science*, 6th edn. Woodhead Publishing, Oxford.

Leistner, L. (1995) Principles and application of hurdle technology. In *New Methods of Food Preservation*, ed. G.W. Gould. Blackie Academic and Professional, London.

Long, L., Komarik, S.L. & Tressler, D.K. (1982) *Food Products Formulary: Vol. I, Meats, Poultry, Fish, Shellfish*. AVI, Westport.

Ockerman, H.W. (1989) *Sausage and Processed Meat Formulations*. Van Nostrand Reinhold, New York.

Ranken, M.D. (1984) *Notes on Meat Products*. Leatherhead Food Research Association, Leatherhead.

Richards, S.P. (1982) Abattoir by-products – their utilisation and investigation of possibilities for inclusion of three offals in meat products. PhD Thesis, Brunel University.

Street, C.A. (1991) *Flour Confectionery Manufacture*. Blackie, Glasgow.

Stumbo, C.R. (1973) *Thermobacteriology in Food Processing*, 2nd edn. Academic Press, London.

Index